Homeowner's Guide to Electrical Wiring

Larry Mueller

Ideals Publishing Corp.
Milwaukee, Wisconsin

Table of Contents

Technical Advisor, Robert Spinti
School of Industry and Technology
University of Wisconsin—Stout

ISBN 0-8249-6108-0

Copyright © 1981 Ideals Publishing Corporation

Published by Ideals Publishing Corporation
11315 Watertown Plank Road
Milwaukee, Wisconsin 53226

Editor, David Schansberg

Photos and drawings by Larry Mueller
unless otherwise indicated.

Cover design by David Schansberg. Materials courtesy of Elm Grove Ace Hardware.

Cover photo by Jerry Koser.

SUCCESSFUL
HOME IMPROVEMENT SERIES

Bathroom Planning and Remodeling
Kitchen Planning and Remodeling
Space Saving Shelves and Built-ins
Finishing Off Additional Rooms
Finding and Fixing the Older Home
Money Saving Home Repair Guide
Homeowner's Guide to Tools
Homeowner's Guide to Electrical Wiring
Homeowner's Guide to Plumbing
Homeowner's Guide to Roofing and Siding
Homeowner's Guide to Fireplaces
Home Plans for the '80s
How to Build Your Own Home

Introduction to Electrical Wiring

Most of us who drive automobiles, and perhaps change oil, pump gasoline and replace spark plugs, have only a vague notion of how internal combustion engines really work. In fact, we go through life working quite competently with a great many highly complex things that we know little or nothing about. Electricity is no different. Nearly all of us can perform repairs around the house, add new circuits, even completely wire a house with almost no knowledge of exactly how electricity performs its miracles in our homes.

Then why do most of us place electricity alongside snakes and spiders in our irrational fears? Maybe because our earliest conditioning toward electricity was a terrifying flash across the sky, followed by earth-shaking thunder that thoroughly convinced us our terror was well founded. But most snakes and spiders, too—are beneficial. And our irrational fears would vanish with just a little knowledge of which ones are dangerous. And so it is with electricity. It works unseen and is capable of being very dangerous. But with just a little bit of understanding it suddenly becomes a safe and easily controlled servant.

Complete Circuits

The one thing you need to know for your own safety and to make electricity work for you is that electricity is not possible unless there is a complete circuit.

Electricity often has been explained through a comparison to water. Water pressure pushes water through a pipe. Electrical pressure pushes "current" through a wire. But that is where the analogy ends. If you open a faucet, water flows out of the pipe. Electrical current can never leave the wires. And it can never flow unless there is a circuit with both a start and a finish.

Electrical current flow is actually the movement of free electrons from one atom to another along a wire. And the electrons don't really flow. It's more on the order of lining up several pool balls and hitting the first one with a cue. The force just bumps from one to another, and only the last ball is set in motion. With electricity, electrical pressure hits the first electron like a cue stick hits a ball, and with the speed of light the force bumps from one electron to the next until it reaches the other end of the circuit. However, if the circuit is open, or not complete, none of this electron movement can take place.

The most simple circuit is a flashlight battery, a bulb, and two wires.

The battery provides the electrical pressure. We call it voltage. The more voltage, the higher the pressure.

A flashlight battery is only 1.5 volts. You would feel the pressure or shock with a 45 volt battery. It would give you a real jolt if the battery had 120 volts like the receptacles in your house. Obviously, for your own safety, you want to recognize how to

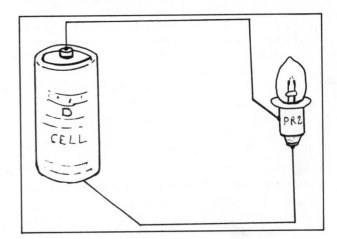

The most simple circuit is a battery, a bulb, and two wires. But learn how it works, and you suddenly understand a great deal about how most electrical circuits work.

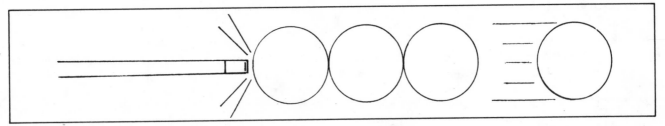

The flow of electrical current is the movement of electrons in a wire. But electrons do not flow like water. They bump and displace one another and pass the energy along much like a row of pool balls when struck with a cue.

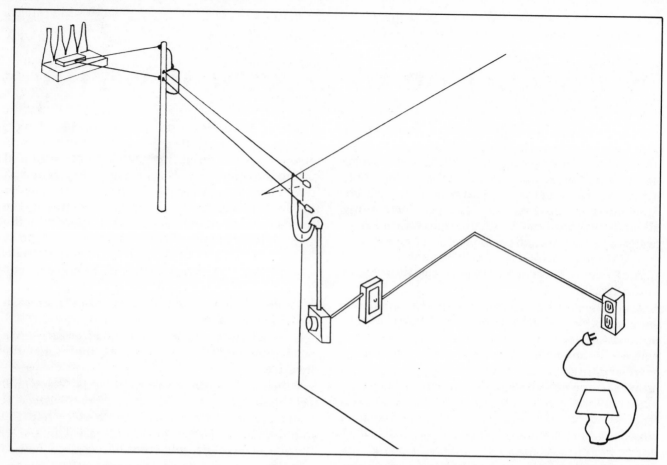

In this oversimplified drawing, we show how electricity is brought by wires from the power company's generators to the transformer on a pole near your home, into the meter, then to the circuit breaker panel, and finally to a wall receptacle where a lamp can be plugged in to com-plete the circuit. While the size, distance, and mechanical means of delivering electricity into the home are quite different, the electrical circuit is just about the same as our simple battery and bulb set-up.

avoid becoming part of a complete circuit.

If we connect a wire to each end of the battery, we now have the electrical means for completing a circuit. Pairs of wires, similar in principle to the two wires on our battery, are running all over our homes, providing the electrical means for completing circuits in house wiring. They don't begin at batteries, of course. They go back to the fuse box or circuit breaker box, and before that, the voltage comes from the transformer on the pole. That voltage comes from the transformer on power lines. And the voltage originates at the powerhouse generator. Always, however, there must be the means for a complete circuit. At least two wires must always be present so "current" can flow both away from and back to the voltage source.

If you touch only one of the wires from the flashlight battery, you will feel nothing. If the battery is 45 volts, and you touch only one wire, you still feel nothing. If you touch only one of the wires which go to a 120 volt socket or receptacle in your home and are not standing or touching a grounded surface, you feel nothing either. You are safe from shock

because you have not completed the circuit.

One way to complete our battery circuit would be to touch the two wires together. We'd get a little spark; and if we kept the wires together, the battery would quickly go dead. If we touch two house wires together, we get a much, much bigger spark and we blow out a fuse or trip a circuit breaker. What we create is a short circuit.

A good practical example is the insulation breaking off of the two wires in a cord going to a lamp. The wires touch and form an easy closed-circuit path for electrons. The current takes this short-cut, and never reaches the lamp. Too much current flows through this short-cut, however, and the fuse blows.

The proper way to complete an electrical circuit is by terminating the two wires with some kind of electrical device. It can be anything from a light bulb to a refrigerator. In our simple circuit, it is a flashlight bulb. When we solder the two wires to it, the bulb begins to glow. It does not turn the battery out as quickly as does a short circuit. The reason? The filament in the bulb offers resistance to the current flow. In fact, it is this resistance to electron movement that causes the filament to heat and glow. This resistance also limits the amount of current that the voltage can push through the circuit. So the battery lasts longer.

If you have actually put together this simple little circuit, you probably used a PR2 bulb out of a 2-cell flashlight. And you noticed that the single battery does not make it glow very brightly. But if we put two 1.5 volt batteries in series (end to end as they are in a flashlight) we double the voltage. And 3 volts can push twice as much current through the bulb as 1.5 volts. There is more electron resistance in the filament, more heat, and a brighter glow. If we add another two batteries in series and get 6 volts, we will promptly burn out the little bulb because of too much current flow. Every electrical device is designed to operate on a certain voltage or within a certain voltage range. Most devices in our homes are designed for 120 volts while things such as stoves, air conditioners, and clothes dryers usually require 240 volts.

Before long, the flashlight battery will grow weak from continuous use. This can be prevented by breaking the circuit. Simply cut one of the wires. And now we have the means to open and close the circuit. The two ends of the cut wire become a simple switch. Touch them together, and the bulb glows. Separate them, and it stops. Switches vary in construction, but all of them do the same thing. They open and close circuits. They are gates which stop or permit current flow.

Chances are, you are becoming impatient with this simple little circuit. You are anxious to get on to the actual house wiring. If you understand this little battery circuit, you already know how an ordinary light circuit works. Only the materials are different. The hook-up is the same. Two wires come from a voltage source to a bulb, perhaps in a ceiling fixture. One wire is broken with a switch on the wall.

You also know how a table lamp is wired. Again, the materials are different, but two wires come from a voltage source. In this case, it is a receptacle. The two wires connect to a socket that holds the bulb, and one wire is broken with a switch.

I did leave out one little detail! The battery is direct current (d.c.), and house wiring operates on alternating current (a.c.). Direct current flows in one direction only. Alternating current reverses direction 60 times every second. But don't let this throw you. If there were a way to reverse the battery 60 times a second without breaking the connections, it would also supply an alternating current of sorts. Either a.c. or d.c. requires a complete circuit to function. And the hook-up is the same with either. Only the devices that are plugged into d.c. are sometimes (not always) different. If you understand a complete circuit for one, you understand it for the other.

If we remove the bulb and touch the two wires together, we have an easy path for large amounts of current flow. The battery will soon be dead. Placing our fingers across the battery did not create a short circuit because our skin offers "resistance" to current flow.

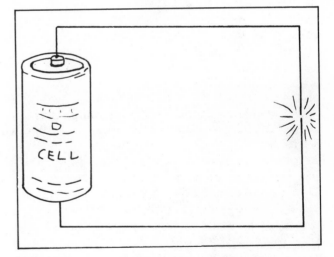

Cutting the wire gives us a crude switch that can open or close the circuit. An open circuit is one in which current cannot flow because the path for electron flow is incomplete.

Electrical Wiring Tools

A continuity tester, sometimes called circuit tester, is one of the handiest tools you can have around the house for electrical repair. Very often a defective electrical device will give no outward physical evidence that anything is wrong. But the tester will quickly reveal its condition.

The continuity tester is never used to test any device that is "hot," that is, plugged into house wiring or any other voltage source such as a battery. The continuity tester is only used on disconnected devices to determine if those devices have open or closed circuits.

A continuity tester is a simple circuit much like the one in the previous chapter, but it has an "alligator" clip and a test probe point instead of the primitive, cut-wire switch.

Continuity means continuous, unbroken. It tests circuits to find out if they are complete or if they are open (broken) somewhere along the line.

When we touched the ends of the cut wire in our simple circuit, that completed the circuit and the bulb glowed. By the same token, and for exactly the same reason, if we touch the alligator clip to the probe point, the bulb glows in the tester. Furthermore, if we place anything between the probe and clip that will conduct electricity, we complete the tester's circuit and the bulb glows.

To test a table lamp, connect the tester across the prongs of the plug. No glow in the tester. Turn the lamp's switch on. Still no glow. Remove the light bulb, and test. It is good. The open circuit has to be in one of the two wires in the cord or in the switch. Remove the socket and switch from the lamp. (Pry off the end cap and pull the socket from its housing.) Test each wire. Good. The outside metal portion of the socket is connected directly to one screw connection. The center contact in the socket is connected through the switch to the other screw connection. Connect the alligator clip to the center contact. Hold the probe to the screw contact. The tester does not light? Turn the switch. Now the tester should light. If it still does not, the switch is open (broken) and must be replaced.

A simple device to check "hot" devices to determine if voltage is present is a test lamp which resembles a little neon bulb with wires connected. The wires are usually terminated in probe points so we can readily touch them to anything we need to test for the presence of voltage.

Let us say a floor lamp is plugged into a receptacle, but it will not light. A new bulb makes no difference. We begin to suspect that maybe the receptacle is dead. So we press one of the test probe points into each of the slots in the receptacle. If the test lamp

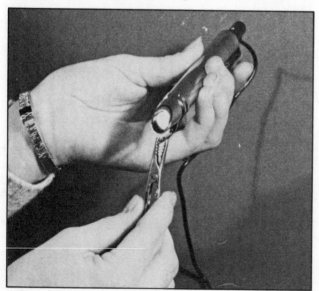

This is a continuity tester. It is just another battery and bulb with mechanical differences. Touching the alligator clip to the probe to light the bulb is not much different than touching two wires of our cut-wire switch.

A test lamp is nothing but a neon lamp with two wires attached. It lights when voltage is present and tells us when the circuit is "dead" by not lighting.

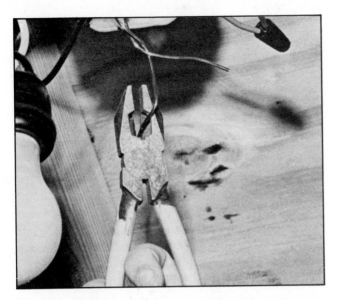

An electrician's pliers, or "sidecutters," are superior to ordinary pliers. It has good, wide, strong-gripping jaws and is an effective wire cutter.

glows, voltage is present in the receptacle, and we can look elsewhere in the lamp for the trouble. If the test lamp does not glow, the next step is to check for a blown fuse or tripped circuit breaker.

These two simple testers, the continuity tester and the test lamp, are all you will need to test any household wiring problems and many small appliance troubles.

Hand Tools

The hand tools you will need are also simple. Most wiring repairs can be accomplished with a screwdriver of appropriate size, a pocket knife to strip insulation from the wires, and an ordinary pair of pliers.

Electrician's pliers, also called "sidecutters," have good wire cutters on one side to the rear of the jaws. And the jaws are great wide wire grippers.

A popular wire cutter is the diagonal cutter. Insulation can also be stripped with this cutter. Insulation strippers will do the job with one pinch of the wire if it is placed in the proper slot. These strippers are combination tools which also act as pliers, cutters, crimpers for certain wire connectors, and wire gauges.

Another handy tool is the needlenose pliers. It too has a wire cutter. But the long, tapering jaws are its main asset. When you need to twist a loop in a wire so it can be held under a screw, nothing does it as quickly and easily as a needlenose pliers.

These simple tools are all that the homeowner needs for occasional electrical repairs. However, if you plan to add circuits, rewire a house, or wire a new house, you will need a few more items.

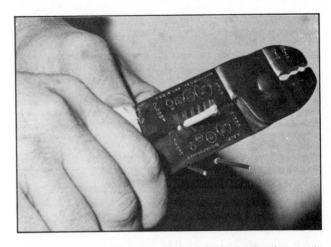

Insulation strippers are combination tools that strip insulation, cut wires, crimp special connectors, and identify wire sizes. To strip insulation, place wire in the proper slot and squeeze the handles together. When the handles are squeezed together and the insulation cut, move the stripper away from the wire quickly, and the insulation pulls off cleanly.

The long, tapering jaws of a needlenose pliers are perfect for bending hooks in wires so they can be held under the screws of switches, receptacles, and sockets.

If you plan extensive work, a sabre saw is extremely useful. It speeds up the jobs of enlarging holes and cutting new openings in walls.

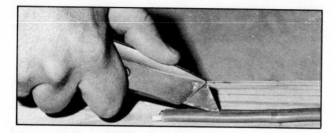

An ordinary pocket knife can remove the sheath from cable, but a utility knife is handier. Lay the end of the wire on a flat board, and draw the knife down the middle of the sheath. After the sheath is split, pull it back and cut it off.

An electric sabre saw is extremely useful for cutting new openings in walls, enlarging holes and cutting wallboard. Hand tool substitutes are keyhole saws (for enlarging holes and cutting new openings) and utility knives (for cutting wallboard).

A drill is essential for rewiring or adding circuits. Electric is fastest and easiest, but you can get by with a hand operated brace and bit. Use a spade bit for going through wood, and a masonry bit for bricks or concrete block.

You will need a hammer for stapling cable and nailing outlet boxes in place.

For pulling cable through walls, conduit or pipe for enclosing wires, a "fish" tape is essential. A fish tape is a springy, flat, metal tape that is worked through the wall first, usually from one outlet hole to another. Then the wire is fastened to the end of the fish tape, and both tape and wire are pulled back through the wall.

A few extra tools are required for installing conduit. Conduit is more expensive and harder to work with than flexible nonmetallic sheathed cable, but if your local electrical code requires it, you have no choice. It may be required only in certain locations, however. One place is on the face of a concrete basement wall where the circuit must drop from the ceiling to provide voltage to a socket, clothes dryer, the furnace, etc. You will need a hacksaw or a pipe cutter to cut thinwall conduit, which is easier to work with than rigid conduit. And a conduit bender will be necessary for shaping the conduit to carry wires around corners.

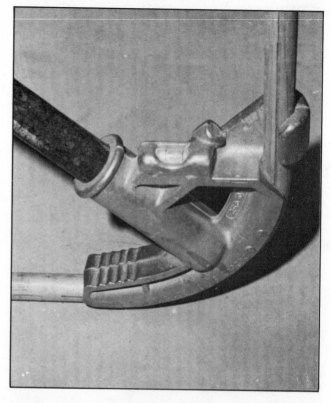

If you need to run conduit around a corner you will require a conduit bender. Trying to do the job without this tool will only get you a bunch of kinked bends.

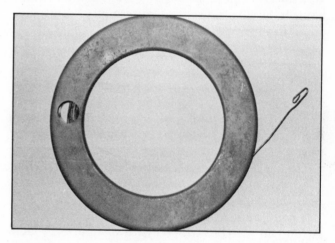

A fish tape is used to snake wires through walls and conduit.

Wires

There are two kinds of wire: copper and aluminum. The most important thing to remember about aluminum is that you do not want it if you have a choice.

Aluminum has only one advantage. It is cheaper. Its disadvantages are many. It expands and contracts more than copper, making loose connections a greater possibility. It corrodes the moment air hits it, so the likelihood of a poor connection increases. Joining two dissimilar metals, such as aluminum to brass, also causes corrosion. And loose or corroded connections on switches and sockets can become so hot they cause fires. On top of that, aluminum wire has to be larger to carry the same current as copper. So, it makes a bulkier wad of wire to be stuffed into the outlet box.

If your home happens to be wired with aluminum, do not panic. About two million are. But be aware that the only switches and sockets safe to use with aluminum are marked CO/ALR (for devices with 15 or 20 ampere current ratings) or CU/AL (for over 20 amperes). If it becomes necessary to replace switches or receptacles, use only those that are so marked.

Also, it would not hurt to periodically check that screw connections on switches and receptacles are very tight. Devices with push-in connectors are never to be used with pure aluminum wire. And while you are at it, place your hand on the cover plates occasionally. You will become accustomed to how warm they normally feel. If one day a plate feels abnormally warm, get the power off to that circuit immediately, remove the plate, and determine whether a fire is in the process of starting. If you have aluminum conductors in your present wiring, any copper wire used for changes or repairs must be connected to terminals and conductors that are labeled CO/ALR or specifically labeled for such use.

Wire comes in various sizes indicated by numbers. The smaller the number the bigger the wire and the more current the wire can safely carry; the bigger the number the smaller the wire and the less current the wire can carry.

In the past, most house wiring was done with number 14 copper. It safely carries 15 amperes of current, and its use is still approved by the National Electrical Code and permitted by some local codes. Over the years, however, we have added more and more electrical conveniences. There has been a tendency to overload our circuits. Overloading is both

COPPER WIRE SIZE	*AMPERES SAFELY CARRIED
14	15
12	20
10	30
8	40
6	55
4	70
2	95
1/0	125
2/0	165
3/0	195

*type of insulation may affect safe current.

dangerous and energy wasteful. The wires become hot, voltage drops, and appliances operate inefficiently. Costly kilowatts are lost in dissipated heat throughout the system, so most modern homes are wired with number 12 copper which safely carries 20 amperes of current.

Larger wire is needed for special circuits. Air conditioners, clothes dryers, dishwashers and other large appliances require number 10 copper. Electric ranges need number 8 or 6 depending upon the current requirements of the particular unit. If in doubt, going to the larger number 6 size is the safe choice.

These wire size numbers reveal the physical diameter of the copper wire. They tell you little about how the wire looks, only its current carrying capacity. When you ask the store clerk for number 12 wire, he may ask, "Do you want that in UF, NMC, AC, or NM type? Two or three wire? With or without ground?"

Don't let this throw you. These code designations just tell you whether the wire can be used underground because of the solid plastic outer covering (UF), whether it is designed for damp indoor circumstances (NMC), whether it has a steel armor (AC) or whether the insulated wires are wrapped with paper inside a plastic sheath and must be used in a dry indoor location (NM).

The common type you will find is the plastic sheathed cable (commonly called Romex), containing at least two wires. As you will remember, to have a complete circuit, one wire must go the light bulb, or whatever device, and a second wire must return. It is easier to pull the wire through walls, etc., if the two wires are encased in a single sheath.

Today, however, it is likely that your local electri-

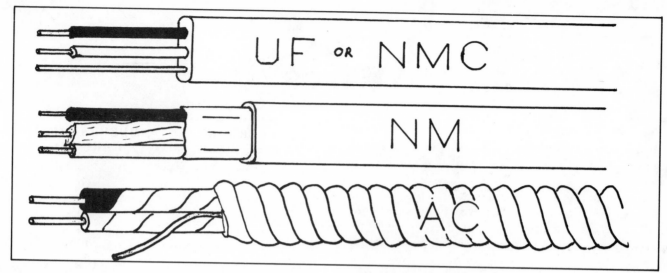

These are the common wire types used in residential wiring. The NM type non-metallic sheathed cable has a paper wrapping between the insulated wires and the outer plastic sheath. It is intended for dry, indoor locations. The wires in both NMC and UF are solidly encased in plastic. The NMC type is intended for damp, *indoor locations. The UF type can be used in damp locations and also is tough enough to be run underground. The AC type is armor covered with a flexible steel winding. It rusts, so it can only be used in dry locations. It is useful in hazardous locations where something may strike and cut other cable types.*

cal code will require you to use "2-wire with ground." That means the cable contains two insulated wires and one bare green colored grounding wire.

What does the bare grounding wire do? Think about that complete circuit again. One wire out, one wire back. Black is hot; white is common or ground. The bare wire goes back to the circuit breaker or fuse panel and connects to the same place the white wire connects: ground. Essentially, we are completing the circuit a second time. It is intended to provide an extra ground for the metal frames of such electrical devices as hand drills.

With that behind us, let us go on to the 3-wire cables and 3-wire with ground. If you are running 120 volt lines through the house, you need 2-wire cable. If you are wiring a receptacle for an electric range, air conditioner, or clothes dryer for 240 volts, you will need 3-wire cable.

Not many local codes require type AC (armor covered cable, commonly called BX) but find out before you begin. It is useful for running through walls where nails might puncture plastic sheath. It rusts, however, and is not useful in damp locations.

The only house wire that is still commonly found in single conductor instead of multiconductor cable is very large service entrance wire: number 3 for 100 ampere service, 3/0 for 200 ampere service.

The following method works best for connecting wires under the screwheads at switches or receptacles.

When you place your screwdriver in the slot of a screw and proceed to tighten it down, you are turning it in a clockwise direction. If a wire is to be held under that screw, curl the wire in the same clock-

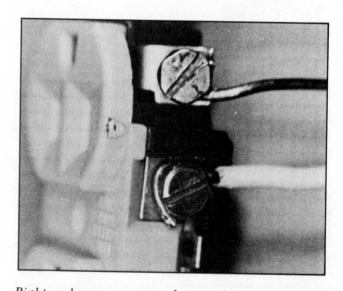

Right and wrong ways to fasten wires under screws. Where the wire is bent clockwise (top wire), the turning screw tightens the bend. Where the wire is bent counterclockwise (bottom wire), the turning screw tends to open the bend, making a poorer connection and exposing some of the bare wire.

wise direction. When the turning screw head grabs the wire, it twists the loop all the tighter. That is the right way.

Joining wires in an outlet box, or junction box, is most easily accomplished by using "wire caps." Choose the proper size for your wire as indicated on the package, bare the wire the recommended distance, hold the two wires parallel, slip on a cap, and simply twist. A good, solid connection is made and insulated faster than it takes to tell.

Plan Now to Avoid Problems Later

You could look at a few pictures, read a few words, grab the tools, and go at it. The circuit probably would work, and chances are you would not even burn the house down. On the other hand, there is a strong chance that what you accomplished would not prove satisfactory and eventually would have to be corrected or added to. The person who plans carefully might get a late start, and he may even finish last, but if you don't have time to do the job right the first time, how will you find time to do it over?

The first intention of making an electrical wiring plan is to end up with safe and adequate electrical service where you need it. Start with a list, or a floor plan, of where you have or will have your various electrical devices. Include notes of their power consumption in watts by checking the specification plates on the appliance. If that is inconvenient, use the following chart of approximate wattages. This will help you plan the circuits and provide adequate electricity where needed.

Codes and Inspections

The National Electrical Code is like a friendly police officer, employed to help us, designed to ensure our safety, and no threat to the average law-abiding citizen.

The National Electrical Code is not law. It is a very complete set of standards recommended for the safety of our property and ourselves by the National Fire Protection Association, 470 Atlantic Avenue, Boston, Massachusetts 02210. The code book is available at this address for $8.25. The book is revised every three years.

The Code book is not a how-to manual. In fact, it is about as much fun to read as government regulations. But it is very useful as a reference book to look up what materials and practices are acceptable for a certain application.

While the National Electrical Code is not law, it does become something that must be obeyed when it is adopted by your local department of building and zoning. Local codes are generally more restrictive. Check with your city or county zoning offices. Local power utilities may also be aware of local codes.

As a general rule, repairs and minor wiring additions can be done by the home owner without a permit. Major rewiring, or wiring a new house, most likely will require a permit from your building and zoning office. At a certain point in the work, an inspector will want to see the job to determine whether you have followed the local codes. Never assume that doing a job according to the National Electrical Code will suit local authorities. Before undertaking any electrical wiring job be sure to check local requirements for permits and inspection.

Codes are intended for your safety. If you follow the instructions in this book and inquire about additionally restrictive local codes, you should be able to do the job right the first time and pass inspection.

TYPICAL WATTAGES OF HOME APPLIANCES	
Appliance	Watts
Air conditioner (room)	800 - 1,500
Air conditioner (central)	3,000 - 7,000
Blanket	150 - 450
Blender	250 - 450
Broiler	1,500
Can opener	100
Coffee pot or maker	600 - 700
Crock pot	75 -150
Deep fryer	1,150 - 1,350
Dehumidifier	400 - 600
Dishwasher	1,200 - 1,800
Drill (hand)	200 - 400
Dryer (clothes)	1,400 - 5,600
Fan	75 - 500
Freezer	350 - 600
Fry pan	1,000 - 1,500
Furnace (electric)	10,000 - 34,000
Furnace (gas or oil)	600 - 1,200
Furnace (heat pump)	10,000 - 33,500
Garbage disposal	300 - 900
Hair dryer	300 - 1,500
Heater (space)	500 - 3,000
Iron	1,000 - 1,100
Microwave oven	650
Mixer	100 - 200
Oven (built-in)	4,500 - 7,800
Radio	10 - 150
Range	8,000 - 16,000
Refrigerator	250 - 350
Rotisserie	1,200 - 1,400
Stereo	300 - 500
Stovetop	5,000 - 6,700
TV	40 - 350
Toaster	1,100 - 1,500
Vacuum cleaner	500 - 600
Waffle iron	1,000 - 1,200
Washer (clothes)	700 - 900
Water heater	2,500 - 5,500

The kitchen generally requires more electricity than any other room in the house. Several appropriate outlets are necessary for the many appliances, small and large, and lighting. Photo courtesy of Armstrong Cork Co.

Kitchen Circuits

Let us start our planning in the kitchen. The electric range and oven will be treated separately, but other than that let us say we have a refrigerator (300 watts), a fry pan (1400 watts), a toaster (1100 watts), a can opener (100 watts), blender (250 watts), coffee pot (600 watts), dishwasher (1800 watts), waffle iron (1100 watts), garbage disposal (900 watts), and a rotisserie (1400). If everything was turned on at once, that would total 8950 watts.

What size wire would that take? First, figure the current flow. Just divide watts by 120 volts. If you divide 8950 watts by 120 volts it equals almost 75 amperes of current. To carry that much current safely we would need number 6 wires.

Obviously, we could not hook huge number 6 wire to wall outlets. If we could, wiring one circuit for that kind of load would require a circuit breaker so huge that trouble in one appliance probably would not trip it. And, of course, we do not operate all of these appliances at once, so there is no need for such a circuit.

The electrical code provides for common sense and safety in the kitchen. Only two double receptacles should be connected to each single circuit of number 12 wire with 20 amp circuit breakers or fuses. If the cook plugs the dishwasher (1800 watts) and the rotisserie (1400 watts) into the same circuit, that is a total of 3200 watts. Divided by 120, that is

over 26 amps current flow, and the 20 amp fuse will blow. He learns to avoid this combination hereafter. But no harm was done because the smaller line with fewer outlets (as opposed to many receptacles on heavy wire) could be safely fused. Furthermore, two double receptacles rarely would have more than one or two small appliances plugged in and turned on at once.

It is up to you, now, to determine how many of these appliances might be plugged in at any one time and where they would be located at that time. This quickly tells you how many circuits are needed in your kitchen and where. The code requires at least two. Lighting may not be connected to these special kitchen circuits.

The receptacles for small kitchen appliances should be placed about 6 inches (152.4 mm) above the counter top and about 36 inches (914.4 mm) apart. The cord of an appliance placed anywhere on the counter should be able to reach a receptacle.

Lighting is on a separate circuit. Lights over the sink and stove are essential to keep the cook from working in a shadow. Another light may be required over the counter, depending upon its location in regard to other lighting. Wall switches, not pull chains, should be used on these lights because of possible shock when one wet hand is in the sink and the other is touching the chain. The ceiling fixture to provide overall light should be controlled from switches on the wall near the door(s).

The kitchen lights add up to 4 or 5 outlets. Since lighting consumes little power, we can plan to add 5 or 6 more outlets to this circuit, perhaps in another room or wherever convenient. The National Electrical Code suggests no more than 10 such outlets on a single circuit. Check your local code. Some are more restrictive.

When planning, consider that the shorter the run of wire, the less chance of energy loss through voltage drop in the wire itself. And the shorter the run, the cheaper the job. Not always, but often, especially when wiring old houses, it is wise to run lines for all receptacles through the basement and up while running overhead lighting and accompanying switches through the attic and down.

In this case, the overhead lighting for two or more rooms might be on one circuit while the 120 volt outlets (receptacles) in these rooms are on another. It is not necessary to fuse circuits separately to individual rooms. Just keep track of all circuits. Make notations on the inside of the door to the circuit breaker panel or fuse box, using the chart provided for that purpose.

Kitchen Range Circuit

Unless you have electrical heating, the electric range is the watt whopper of your home. The average range, with all burners on, will spin the meter at the rate of 12,000 watts, or 12 kilowatts. If we operated this range on 120 volts, it would draw 100 amperes (12,000 watts divided by 120 volts equals 100 amps).

But the range does not use 120 volts. It requires 240 volts. And an interesting thing takes place when we double the voltage. We still need the same wattage, or power consumption, to get the required amount of heat, so when we divide 12,000 watts by

This is a 50-amp kitchen range receptacle. Note the 2-screw connectors that hold the large number 6 wire.

240 watts, we are now using 50 amperes of current. A wire big enough to carry 100 amps would have been unmanageable. A number 6 wire can carry the 50 amps.

Plan on running one 3-wire number 6 cable to operate the range. Nothing else may be on this circuit. Protect it with a two-pole 50 amp circuit breaker. Terminate the circuit with a special 50 amp receptacle to take the special range plug.

If you are installing a built-in oven and stovetop (separate units) it may be easier to run a different circuit for each. The cables for these circuits would be 3-wire number 10. If that does not seem like large enough wire, remember that half as much current is required for a given wattage when the circuit uses 240 volts instead of 120. Each number 10 wire 240 volt circuit should be protected by a two-pole 30 amp circuit breaker.

If you want to use receptacles and plugs so the units can be easily disconnected for service, special 30 amp receptacles and pigtails (cords) with plugs are available. The wire in the pigtails must also be number 10.

Ovens and stovetops do not require frequent service, and complete removal of the built-in units will be rare. If you would rather not be bothered by the trouble and expense of receptacles and pigtails, it is all right to connect the number 10 circuit directly to the wires on the units. Further details can be found in the section on actual wiring.

Dishwasher and Laundry Circuits

Dishwashers require from 1200 to 1800 watts and operate on 120 volts. Two-wire with ground, number 12, is adequate. But to meet the requirements for the National Electrical Code, give the dishwasher its own separate circuit. This requirement also applies to garbage disposals and any other appliance which is rated at 12 amps or more and is not readily portable.

Laundry appliances might be installed in the kitchen, the basement, or in a utility room. If you choose the basement, plan for the circuit "drop" (wire coming from the ceiling down the concrete wall to the receptacle) to run in conduit. This holds the cable solidly against the wall and provides additional protection from moisture.

An automatic washer uses 700 to 900 watts. Since this is less than 10 amps, the washer can operate when plugged into an ordinary receptacle rated for 20 amperes. The National Electrical Code calls for one such receptacle on a separate circuit in the laundry. This can be used for the washer.

Clothes dryers use about 5000 watts. Most operate

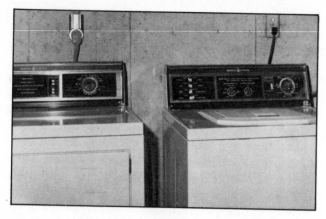

The Code requires one 20-amp circuit in the laundry which terminates in a single duplex receptacle. This can be used for the washing machine. A separate 30-amp circuit (left) is required by electric clothes dryers.

Thirty-amp receptacles can be used for separate stovetop and built-in oven circuits. The same type receptacle is used in clothes dryer circuits.

on 240 volts. Plan to use a 3-wire number 10 circuit protected by two-pole 30 amp circuit breaker. Nothing else may be connected to this circuit.

Electric water heaters are generally used where gas is not available. They are more costly to operate, and the economics of this is simple. Gas or some other fuel is burned to make steam which turns generators to make electricity which eventually heats water. There is an energy loss in every one of those steps—fuel to steam to electricity to hot water—plus the cost of investment and manpower all along the line. It simply costs less to burn gas and let it heat the water directly.

Another disadvantage in using electrical water heaters is their slow recovery rate. Recovery rate depends upon the wattage of the heating elements, of course. As an example, one 30 gallon (113.55 L) tank with a 3800 watt element recovers 17.3 gallons (65.4 L) per hour at 90 degrees Fahrenheit (50 C) rise. And a gas heater of the same size, type of insulation, etc., recovers 34.8 gallons (131.7 L) per hour. As a result, a larger electrical water heater tank is generally required so everybody will have hot water for a bath, or so there will be hot water for every load of clothes in the washer. And this larger tank of water must be kept hot all day, which costs more than keeping a small tank hot.

Whatever kind of water heater you decide on, plan to install it as close as possible to the laundry, bathroom, and kitchen. Avoid long runs of hot water pipe because every time you turn on a hot water faucet, you run the cold water out of the pipes before warm water arrives. While that is happening, an equal amount of cold water is running into the water heater. It has to be heated in addition to the amount of hot water you finally remove from the tank before you turn off the faucet.

Various size electric water heaters consume between 2500 and 5500 watts. Most, but certainly not all, operate on 240 volts. Plan a single 240 volt circuit for the heater (unless it is a 120 volt unit, of course), protected by a two-pole circuit breaker. Although the heater will not use 30 amps, it is still a good idea to use number 10 wire. It will be certain to pass inspection. And there won't be much voltage drop in the line to make the heater operate less efficiently.

If the heater is small, use a smaller circuit breaker. The circuit should not be fused higher than 150 percent of the heater's current rating. A 2500 watt element, for example, will use 2500 divided by 240 volts equals roughly 10 amps. Provide 15 amp protection.

It is also a good idea to check with your power company concerning additional requirements for installing electrical hot water heaters. For example, an electrical timer may have to be installed to shut off the heater during periods of the day when power companies experience peak loads. The idea here is to reduce this peak load so bigger and more expensive generating equipment is not required to meet increased demand during just a short part of the day.

Circuits In Other Areas

Living Rooms The National Electrical Code requires that in all living areas (not the attic and basement) electrical outlets must be spaced so any lamp or other device along the wall is within 6 feet (1.8 meters) of an outlet. These outlets must not be more than 12 feet (3.6 meters) apart. Considering the possibility of tripping over extension cords, the danger of overloading them, and the fact that many devices do not have cords 6 feet (1.8 meters) long or longer, it makes sense to provide more outlets than basic minimum requirements call for.

Usually, outlets are evenly spaced around the room, but this is not a hard fast rule. Plan where the furniture will be. Arrange the outlets so at least some will be exposed. It is annoying to climb around furniture to plug in a vacuum cleaner.

Generally, a number 12 wire circuit in living areas other than the kitchen can safely supply current to 8 to 10 receptacles, or perhaps a combination of 8 to 10 overhead lights and receptacles. If your plans call for a cluster of high wattage devices, you may want to put fewer than 8 outlets on that circuit. Remember that number 12 wire circuits are protected by 20 amp fuses or circuit breakers. A safe total sum of the wattages on any one circuit cannot exceed 2400 watts (20 amps x 120 volts). Simply add up the wattages of the lamps, TV, stereo, etc., that you plan to operate on a single circuit. Figured on a floor plan basis, you should have a 20 amp circuit for every 500 square feet (46 square meters) of living space.

If a window air conditioner will be installed in the living room, plan a separate circuit. Make the run as short as possible to prevent unnecessary voltage drop. Window air conditioners range from 750 to 1500 watts and are safely supplied with current by a number 12 circuit. If it is a 120 volt unit, run 2-wire with ground. If it is a 240 air conditioner, a 3-wire circuit will be required.

Living room lighting usually consists of one or two ceiling fixtures for overall lighting, and floor and table lamps for limited lighting around certain chairs, end tables, etc. Plan switches at all entrances to control overhead lamps. This is convenient, and it saves electricity. People leaving the room through one door may not always walk across the room to another door to turn out the light.

Indirect lighting (hidden lamps aimed at walls and/or ceiling to reflect light back into the room) had a period of popularity, but it is a highly wasteful system. In theory, light doesn't diminish linearly (that is, half as much light when twice as far away); it diminishes with the square of the distance. There is one-fourth the light two feet away as there is at one foot. Bounced or indirect light is glare-free and diffused but expensive, not only because of the dis-

The living room usually relies on lamps for most of its lighting, but track or recessed ceiling lamps can high- *light various areas of the room. Photo courtesy of Armstrong Cork Co.*

tance it travels, but because some of the light is absorbed by the walls. White walls reflect about 80 percent of the light, but medium green walls, for example, absorb about 70 percent of the light, reflecting only 30 percent into the room. The darker the walls (this includes wood paneling), the greater the light absorption.

If it doesn't conflict with the decorative scheme of the house, try to use as much fluorescent lighting in your plan as possible. For watts consumed, fluorescent lamps produce two to four times as much light as incandescent bulbs. And they may last 20 times as long.

Decorative lighting may either be plugged into receptacles or permanently wired and controlled by wall switches.

Family Rooms In some homes, the living room is treated more like the old-time parlor (mostly for company), and a family room becomes the real "living" room. In that case, many of the items otherwise kept in the living room (TV, stereo, etc.) will be installed in the family room, plus others such as popcorn poppers and movie projectors, and perhaps hobby tools such as drills, soldering irons, etc. This room needs an abundance of outlets.

Bedrooms Plan enough outlets for lamps near the bed, electric blankets, clock radio, or electric alarm clock, heating pad, and vaporizer. If a room air conditioner will be in the bedroom, it is best to provide a separate circuit. Overhead lighting might be controlled with a dimmer switch if it will be turned on during the night when the eyes are not used to the light. Dimmer switches which screw into the socket ahead of the bulb are great for bedside lamps.

Bathrooms Provide no receptacles near the tub. The easiest way to get electrocuted is to plug in a heater or radio, get in the tub, and accidentally knock the appliance into the water with you. If an electric heater is needed for comfort, it should be a permanent installation, switch controlled, so it cannot be moved close to a tub or basin.

Lighting consists of an overhead fixture controlled by a switch at the door, plenty of light at the mirror (both sides are better than top lighting), and perhaps a vaporproof lamp in the shower with the switch beyond the reach of anyone in the shower.

Closets If the lamp can be mounted just above the door, and there is no chance of it touching clothes and starting a fire, a pull chain fixture can be used. If the lamp will be out of reach, safely on the ceiling, operate it with a wall switch or special switch that turns the light on when the door is opened. The National Electrical Code requires the lamp to be 18 inches (457.2 mm) from anything that might burn.

Halls Long halls should have switches at both ends to operate the overhead light. If the hall light is at the bottom of the stairs, it is a good idea to use one 3-way switch just inside the entrance and another at the top of the stairs. A second light controlled by the same switches should be installed at the top of the stairs.

Basements A simple, seldom-used basement needs lights at the stairs and enough lamps to light up the room or rooms. If the laundry is in the basement, check the requirements listed under laundry circuits in this chapter. The furnace blower is on a separate circuit. If a workshop is in the basement, provide plenty of receptacles and adequate light. Basement family or recreation rooms should be treated like family rooms.

If a heat pump or resistance heat type electric furnace will be a part of the household wiring, check with the local power utility for requirements.

Attics Good stairway lighting and a bulb or two in the attic is adequate. If it is a large attic and will be converted to a room or rooms later, include a circuit for an adequate number of outlets to be added during the construction work.

Basements do not require fancy lighting, but this fixture was installed with the future in mind. By mounting the box on a hanger between joists (instead of using a surface socket) a different fixture can be installed if the basement is converted into a family room someday.

Planning Extra Circuits and Rewiring

Wiring new construction is easy. It is done before all the studs, joists, and possibly rafters are covered by wallboard. Nail the outlet boxes in place, drill holes, pull the wire through, staple it down, and the rough wiring is about completed. Rewiring older homes is not so easy. Holes have to be made to accommodate outlet boxes and to pull wire through, but you don't want to tear up the whole wall doing it. In some cases lots of advance planning is necessary.

General Planning for Older Homes

Homes built before 1930 generally had "knob and tube" wiring. This means single insulated wires held in place and away from surfaces by two-piece porcelain "knobs" with nails down the centers. Where these single wires went through studs, etc., a hole was drilled first, a porcelain "tube" was inserted, and then the wire passed through the tube.

Check the insulation on these wires at outlets, in the attic, or wherever accessible. If it is crumbling, the wires should be replaced.

Do not add circuits directly to this old knob and tube wiring. It was installed with number 14 wire back when fewer electrical devices were in use. It was common to permit two or three circuits to supply the whole house. Usually, these circuits became badly overloaded with time. It certainly would not be safe to overload them more. Any added circuits should be new circuits starting at the fuse box.

Plan new circuits. Decide where you want fix-

tures, switches, and receptacles. Then study the construction of the house. How will you get the wire from one point to another? Usually, it is not possible to make direct runs as you would in new construction. You will find few shortcuts. Most of the runs will be longer than you like, but often there are not many choices. Always try to keep wire lengths as short as possible.

It may seem impossible to get a wire from the circuit breaker panel in the basement to the attic. But check the soil pipe. It runs from the basement through the roof. It may be possible to run a fish tape alongside the pipe from attic to basement and haul the wire back through with the tape. If there is difficulty getting the tape through because of a tight fit, enlarge the pipe's opening at that particular place. It may be accessible enough for this. If getting the tape alongside the pipe and pulling the wire the entire distance is too difficult, take it one floor at a time. Find where the soil pipe comes through the floor in a closet, or wherever, and go to work.

If the attic has no flooring, it will probably be easy to run wires through and along joists to openings cut for ceiling lights in the rooms below. The wires to switches will have to drop down through the walls.

Examples of the old "knob and tube" wiring. If it has been overloaded, the insulation is brittle and falling off and needs to be replaced. If it is in good shape there is no need to replace it.

Inside Walls

Inside walls are generally the easiest to work with. Outside walls may be solid masonry or filled with insulation. In addition, diagonal bracing and 2 x 4 (38 x 89 mm) cross members to retard the rapid rise of flame are often found in outside walls.

Inside walls of stud wall construction are usually hollow from the floor to the ceilings. At the floor, however, there are 2 x 4 (38 x 89 mm) plates. The upright studs are nailed perpendicular to these plates. At the ceiling, the studs are nailed to "top plates," usually two thicknesses of 2 x 4's (38 x 89 mm). It is easy to fish a wire through the broad open spaces between the studs. But holes have to be drilled through the plates.

Generally, it is not difficult to locate these plates. Plan to drill down through the top plate to drop wires to switches. And plan to drill up from the basement through the bottom plate to run wires to receptacles. But look closely at the situation in your house. It may be necessary to do it differently. In the planning stage all you can do is study all the possibilities and determine the most likely wire routes.

Some very old homes may have brick or stone walls, even on the inside. If the masonry is covered by plaster, it is possible to lay the wire in channels or grooves cut in the plaster. Holes deep enough to accept outlet boxes are chiseled into the masonry. The wire and boxes are held in place by filling the channels and the parts of holes not occupied by boxes with plaster. If this is done, it has to be followed by wallpapering or painting.

Many people prefer masonry walls to plaster. Removing the plaster entirely to enjoy the brick or stone walls leaves little choice but to run wires for switches alongside doors and covering them with strips of wood stained or painted to match the doors. Wood can also hold the switch boxes.

The tube part of knob and tube wiring protects and insulates wires going through walls. If this wiring is kept, do not add extra circuits onto it. Use it only for light load demands.

Another possibility is surface wiring, always an option for older homes where running wire through walls is difficult.

Installing receptacle boxes in masonry is equally difficult. Use a suitable length of plank, fastened to the wall and extending up from the baseboard to hold the receptacle about 12 inches (304.8 mm) above the floor. If the outside walls will be insulated with styrofoam sheets and covered with plasterboard, outlet boxes can be installed on the furring strips.

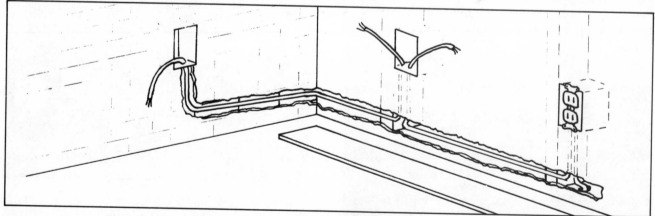

This illustration shows how to deal with two different situations in adding wiring. The already wired receptacle is at the right in the stud wall. To avoid defacing the wall, remove the baseboard and chip away the plasterboard or plaster and run the wire to the next box. The outside wall is plaster-covered brick. There is no solution except to cut a groove in the plaster, lay the wire in the groove, and plaster over it. A hole for the box will have to be chiseled into the brick.

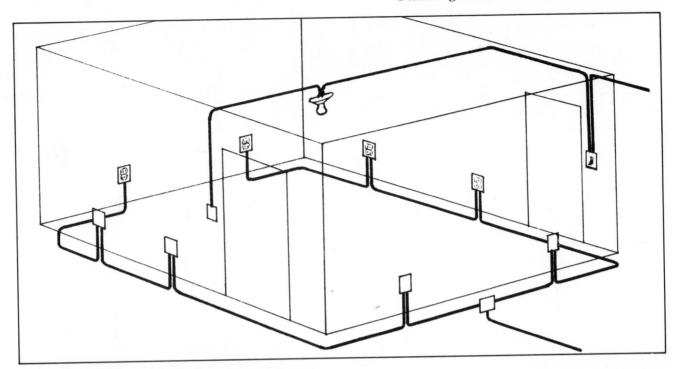

In older homes, it is usually easiest to run cable into the attic and then down to overhead lights and switches. Receptacles are usually easiest to wire from the basement.

This new construction shows the customary two "plates" at the top. Your walls will hide this from your view, but it is there. Wires can run from the attic into the wall through holes drilled down through the plates. Locate the plates by measuring from something such as a soil pipe, stairs, chimney, etc., which is common to both the first floor and the attic. Then carefully drill a test hole. If no plate is felt, stop before going through the ceiling. Plates are generally exposed in attics which are not floored, and therefore are easy to find.

This wire was brought down from upstairs by means of a length of flooring cut away for access. Plaster had been on the brick when the original wiring was done in the 1940s, so a groove was cut for the wire, then plastered over again. The wall will be left brick now, so the rewiring job will be covered with a length of pine plank.

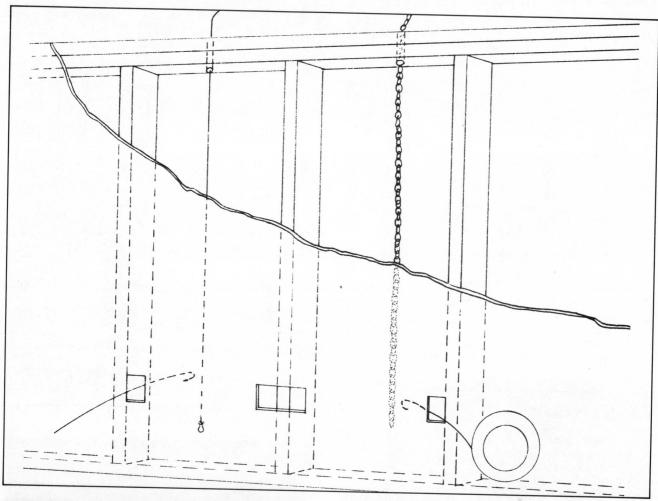

To pull the wire through the wall with the fish tape, remove about two inches (50.8 mm) of insulation and wrap the bare wires tightly around the hook of the fish tape. Wrap it with tape all the way over the hook and back again to the starting point shown here. This will prevent the wire from pulling off the hook and will allow the hook to slide through the wall without catching on something.

This illustration shows the complete studwall construction and three ways to get wires through it without tearing the wall apart. Drill down through the two top plates to gain access. On the left we have dropped a fishing weight on strong cord and will probe until we catch it with a wire hook. Once the cord is through the hole for the box, it can also pull the wire through.

On the right we have dropped a chain through the hole and are probing for it with the fish tape. The tape will be pulled up through the wall, and it will in turn pull the wire back through.

Different combinations also can be used: two fish tapes, a chain and wire hook, etc.

In an apartment situation, or any other circumstance where access is not available through an attic, there is no choice but to cut holes in the wall as shown at the middle stud. With this access it is possible to drill through the stud to run the wire. It will have to be done at each stud, of course, all the way from the existing receptacle to the new one.

To locate the studs, tap on the wall lightly with a hammer or your knuckles. Between studs, the sound will be hollow. You will recognize the change when you strike the solid backing of the stud. There is also a small instrument available in some hardware stores which indicates metal and locates nails which are in the stud you are searching for. Drill a small test hole to be sure you have the stud before cutting large holes in the plasterboard.

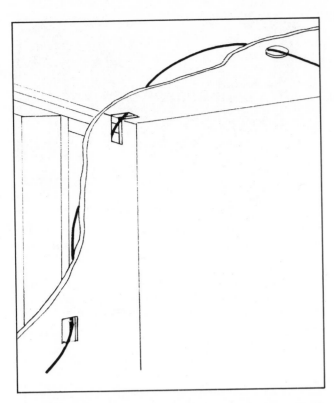

This is the bottom part of studwall construction. Drill up from the basement to run wires into the wall.

In an apartment situation where no access is available from the attic or basement, there is no choice except to cut holes in the plaster or plasterboard. This illustration shows how to get from a hole for a ceiling fixture down the wall to a hole for a switch box. Where the wire goes over the two top studwall plates, secure it with staples before covering it with patching plaster. If the plasterboard is thin, it may be necessary to cut a groove in the plates to get the wire recessed so it can be adequately covered with patching.

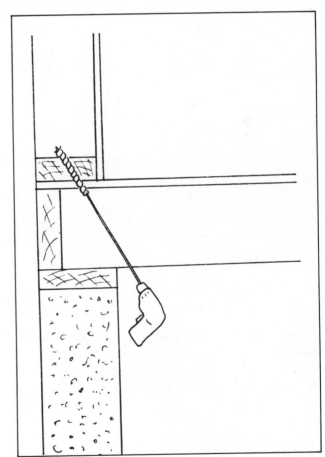

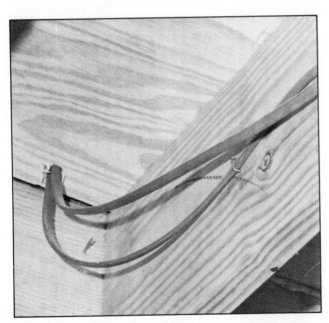

Wires going up from the basement into a wall may look like this. You have to find the bottom plate of the studwall by measuring. It will not be exposed. Plates and studs may be exposed near plumbing or other wiring.

This is a cross-section view at the junction of the foundation and the outside wall. It shows you how to drill through the bottom plate of the studwall to run wire up from the basement.

Electrical Heating

Electrical heating has been encouraged by some power companies, and in some areas it has become common.

Resistive heat is heat derived from an element that becomes hot as current passes through it. Other examples of resistive heat devices are portable heaters and toasters.

Some home resistance heaters are baseboard mounted. This provides the versatility of each room being on a separate thermostat.

An economical form of electrical heat is the heat pump. It works in conjunction with a resistance heat furnace by drawing some of the heat from the outside air. Good units can extract heat from outside air as cold as zero degrees Fahrenheit (–18°C). The result can be almost three times more heat for your money than with straight resistive heat.

If you are planning electrical heat of any type, contact your power company for local requirements. You may need a considerably larger capacity ser-

An efficient heat pump can draw heat from the outside air in temperatures as low as zero degrees Fahrenheit (-17.78 C).

vice entrance to accommodate electrical heating.

Planning the circuit or circuits for electrical heat varies widely according to the type chosen. The unit's specifications, plus local power company and building code requirements, will determine wire size and manner of installation.

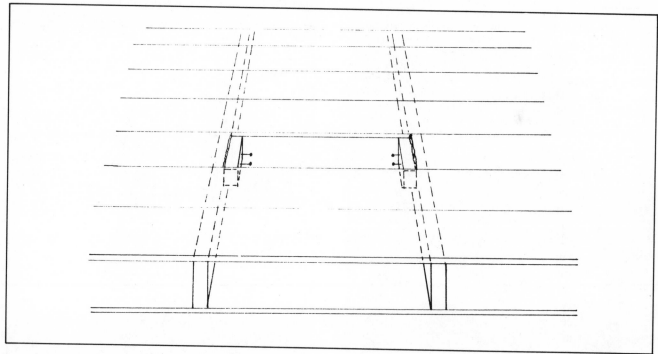

Sometimes there is no attic from which to bring down a wire (or at least it has flooring) and it is not possible or wise to make extra holes in the walls. The only choice is to remove a length of upstairs flooring from between two joists to make access for pulling wires or mounting a ceiling box.

Nails in rough flooring will show you where the joists are, but they will not show in tongue and groove flooring. In this case, tap and listen for hollow sounds between joists and solid sounds when you reach a joist. Drill a small test hole, and then a larger one through the flooring at a point just inside the joist. The larger hole should be just big enough to permit entry of your sabre

saw blade. Saw through the floor board. Repeat this at the next joist.

Pry loose the short length of flooring you have cut free. If the flooring is tongue and groove, you may have to chisel off the tongue on one side before it will come free. A sharpened putty knife is a better tool for this than a regular chisel because the blade is thin and will not make dents in the flooring.

When the wiring job is completed, nail blocks of wood inside the joists as shown. Then you can nail the short length of flooring back in place. The cuts will show, of course, but it is usually possible to cover the whole affair with a rug.

Adding Extra Circuits

It is time for a little more understanding of how electricity works. Remember our little circuit at the beginning of this book? A battery, two wires, and a bulb? How do you add a circuit to the battery for another bulb? Do you wire another in tandem (series) with the original bulb? No. That doubles the resistance to current flow, and both bulbs will be dim. Wire the added circuit across, or in parallel, with the original bulb. Now both bulbs will light. You can keep adding bulbs in parallel, and all will light equally.

Check the accompanying illustration to be sure you understand the hook-up. A second illustration shows how the connections would be made to allow each bulb to be switched on or off separately. We have two wires from a voltage source. All our working circuits are hooked to them. And this is basically how our homes are wired.

Let us carry that one step further. Substitute your circuit breaker panel for the battery. If you want to, you can carry the comparison all the way back to the generator at the power company. The voltage originates there and is carried on wires to your house.

Just as with the battery circuit, two wires are connected to voltage in the panel. These two wires are now encased in plastic or metal sheathing, so their physical appearance is different; but in electrical principle, in the matter of electrical connections, they are quite like the two wires on the battery.

This pair of wires may travel halfway across the house. And somewhere along the line we may have a ceiling light and a switch connected across them. Again, the physical appearances are different, but as far as electrical connections go, the light and switch are quite like the switch and bulb in the battery circuit.

Next let us hook up a receptacle. Electrically, nothing has changed. You will notice two slots in the receptacle. Inside those slots are metal strips. Physically, they look quite different than a pair of wires. Electrically, however, they are extensions of those two wires.

Plug a lamp into that socket. The two prongs of the plug make electrical connections with the metal pieces in the receptacle. Voltage is now connected to the lamp. Current will flow if the switch is turned on. Electrically, the lamp is essentially the same circuit we have in a wall switch and overhead light.

Our basic circuit was a battery and a bulb. But now we need to add a circuit for another bulb in parallel as shown.

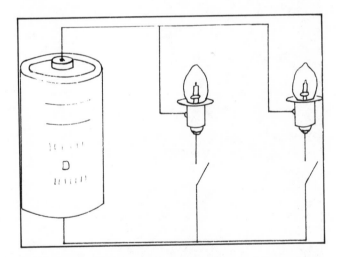

If we want to switch each circuit off and on separately, we add a switch to each leg. House voltage will come from a circuit breaker or fuse instead of a battery, and the switches will be physically different; but electrically, this circuit is basically the same as any light in our home.

And both are electrically connected the same as our bulb and switch in the battery circuit.

What do we do to add another circuit? Connect two wires across the original, run those wires to the new outlet, connect them to the receptacle, and voltage has been made available where it is needed.

Does anything limit how many extra circuits can be added to the original? Yes.
1. The size of the wire in the original circuit.
2. The number of devices already hooked to the circuit.
3. The wattage of the device planned to be connected to the newly added circuit.

Number 12 wire carries 20 amps. Most older homes were wired with number 14 wire which safely carries 15 amps. If there are several receptacles in this circuit already, and if the total wattage of the devices plugged into this circuit approaches 1800 (120 volts x 15 amps), it is not safe to add more power consumption to that line. If it is a 20 amp line, it can safely handle up to 2400 watts. In practice, however, do not push the circuit too close to its wattage limits, or you will be changing fuses or resetting circuit breakers on a regular basis.

Check other circuits. Turn all the lights on in the house. Unscrew fuses, or switch off circuit breakers to see which circuits feed what. (If you are lucky, there is a chart inside the panel door which shows the wiring plan, and you will not have to do this.)

Try to find a circuit that isn't overloaded. If all circuits are operating near capacity, you will have to go back to where additional current is available: the fuse box or circuit breaker panel. Look for an unused breaker or fuse and begin the new circuit at that connection.

If all fused connections are already used, a new panel will have to be added.

Before connecting an additional panel or fuse box, however, be sure to add up the wattages of all devices that may be turned on at any given time. Make certain that your service entrance is capable of handling additional current. Some older homes have only 60 amp, 120 volt service. Others may have 60 amp, 120/240 volt service. These are inadequate for modern wiring and the energy consuming appliances we are accustomed to.

If service is inadequate for your needs, call the power company and arrange to have it increased.

To better illustrate the function of a receptacle, we have broken off the metal hanger and insulating cover. We can now see that the screws fit into metal pieces which serve to conduct electricity to whatever devices will be connected to them.

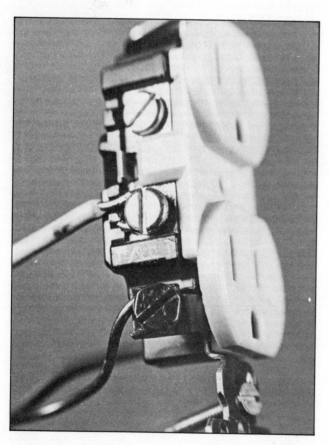

A receptacle looks sort of complicated, but electrically it is nothing more than an extension of the wires connected to it and a means of easily connecting or disconnecting any device to the voltage on those wires by means of a plug. Always connect the white wire under the silver colored or nickel-plated screw, the black wire to the brass screw, and the bare (or green) grounding wire to the green screw.

The same broken-away receptacle now has a plug inserted. It is easy to see how current can flow from the wire to the screw, to the metal, to the prong of the plug, and into a connected device.

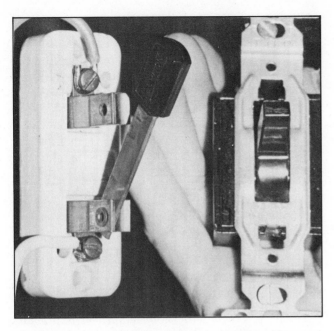

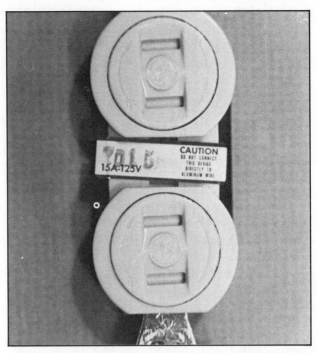

You learned in the beginning of this book that the knife switch looks different than our cut-wire switch, but it does the same thing electrically. The switches commonly used in our homes also look different. They are insulated and box mounted, but electrically they are just like the knife switch with contacts being opened or closed to open or close a circuit.

Electrically, this receptacle is the same as any other. But you may want to use it if you are wiring a small child's room. The prongs of a plug must be inserted in the slots and turned 90 degrees before it can be pushed in to make contact.

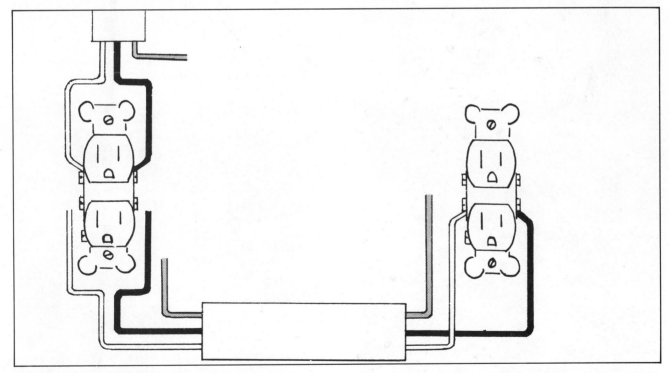

When adding a circuit by connecting to a receptacle, try to use the "end of the run" receptacle. It has just one pair of wires hooked to it. We can install the added receptacle anywhere it is convenient, run our cable to the end of run receptacle, and hook up the wires. This illustration shows the added receptacle about to be connected—black wire to black wire, white to white—providing, of course, that the original receptacle was properly wired with the white wire to the silver colored screw, and black to the brass screw. For simplicity we have not connected the bare grounding wires. Each will go under a screw in the metal box or in a grounding clip on the box. In addition, a short length of bare wire also will be connected from the green grounding screw on the receptacle to a grounding clip or screw on the box.

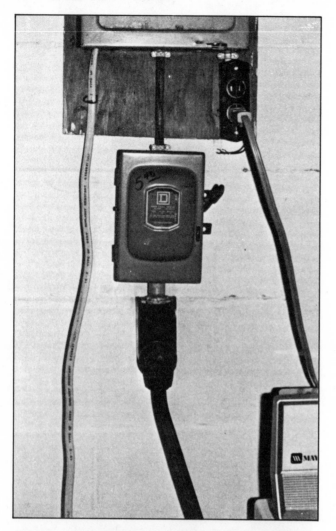

This device, shown with the fuse uncovered, provides additional safety to a leg of the circuit. In this case, the main circuit is fused at 20 amps, but the appliance connected is safer fused at 15 amps. If trouble occurs the 15 amp fuse blows. The 20 amp may or may not blow depending on what else is turned on at the moment. Either way, the appliance has 15 amp protection.

A single receptacle laundry circuit was added here to conform to Code, and a sub panel was installed to provide 240 volts to a clothes dryer.

On the right of this close-up photo is a grounding wire passed through a grounding clip which is about to be shoved in place by a screwdriver. To its left is a grounding clip already in place with the excess wire cut off.

If all existing circuits are operating near capacity, it is not safe to add more. Go back to the panel and look for an unused breaker or fuse. If there is none, disconnect the main fuse or breaker. Hook a short wire to the now cold side of the main fuses or breaker, and connect it to a smaller panel. In this case an extra 240 volt circuit was needed, so a fuse block was added as a sub panel. Be certain when hooking up a sub panel, however, that your service entrance is large enough to supply the additional power required by the new circuit.

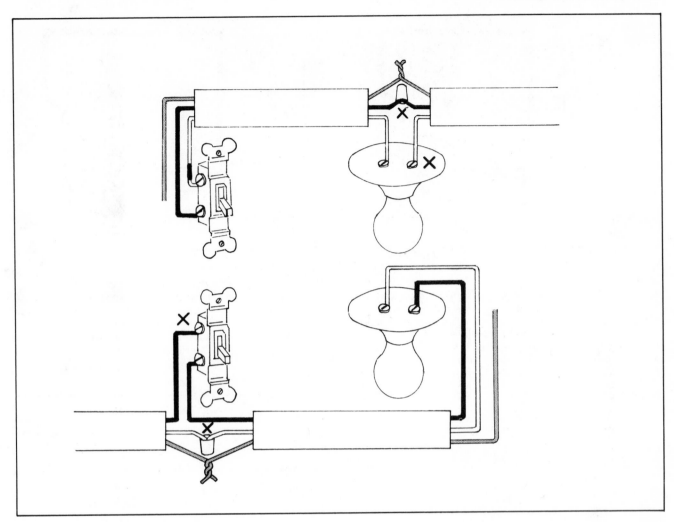

It may be more convenient to add a circuit to an existing lighting circuit instead of to a receptacle. There are two ways to wire single-pole switches and lights. At the top we have the "hot" cable coming to the light box first. The black wire passes through the box by means of a splice and goes on to the switch. We can add a receptacle or another lighting circuit by making our connections at the points marked by X's. Obviously, in this case, we cannot make our connections at the switch box. This is usually evident when we uncover the switch and find a white wire marked with black tape. In the second circuit (bottom) the hot cable goes to the switch box first. We find two black wires on the switch, indicating that the hot wire (black) is broken by the switch before going to the light. Make the connections at the X's. In this case, if wiring has been done properly, finding a black and white wire on the light socket is a tip-off that we cannot make our connections at this point.

Ground Fault Interrupters

The National Electrical Code requires "ground fault interrupters" in all new outside circuits as well as in bathroom and garage circuits. The "GFI" is a very sophisticated type of circuit breaker. Think of our old complete circuit—one wire out, another back. The current that flows out on one wire comes back on the other in an equal amount. But what if a high resistance path (high resistance short circuit) develops to the frame of a hand drill being used outside, for example. If we pick it up to use it, we can get a severe shock. But also a charge occurs in the current balance. Ordinary current to run the drill goes out one wire and in the other. But a tiny amount of current also flows from the hot wire through the high resistance short to the frame and into the bare grounding wire in the cable. Now the hot or black wire is carrying slightly more current—ordinary running current, plus the current to the grounding wire. The white wire carries only the ordinary running current.

The GFI is designed to detect a current difference of as little as .005 of an ampere. And when such a small difference is detected, it interrupts the circuit within 1/40th of a second—too fast to allow you to get a severe shock.

GFI's are two things: great protection and comparatively expensive. Check to see if they are required by local codes. If not required, the decision whether to opt for economy or greater safety is yours.

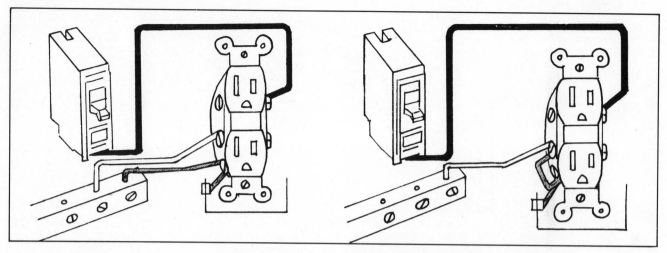

Grounding wires are a necessity wherever ground fault interruptors (GFI) are used. The National Electrical Code recommends GFIs in bathroom, garage, and outdoor receptacles. In other applications, however, the third grounding wire is a waste of money and metal resources. That extra wire is connected directly to the white neutral wire at the neutral bus bar in the circuit breaker panel. The grounding wire does not have to travel all that way to make the connection. It can be connected to the white wire right in the box. And one third of the copper wire will be saved in the process. The drawing at the left shows how we have been hooking up the ground wire, and how it must be hooked up if GFIs are used. The drawing at the right shows how the same thing can be accomplished with a short piece of wire from the white wire connection on the receptacle to the grounding connection. This is the way to go if permitted by local codes.

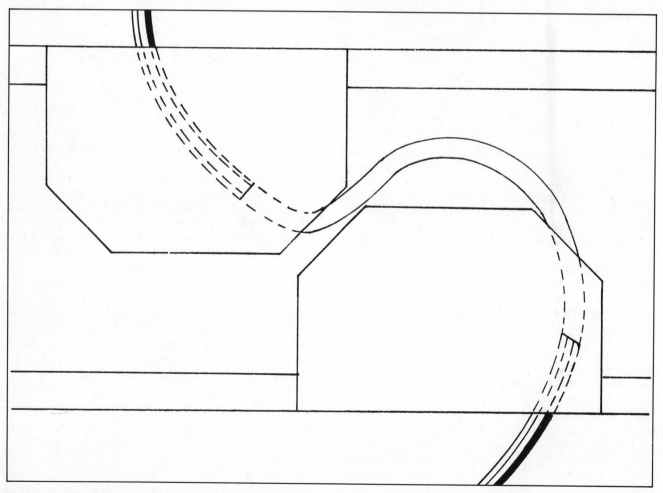

The most simple way to add a receptacle is to place it directly opposite an existing receptacle on the other side of the wall in the next room. A very short length of wire is needed, and fishing it through the wall is little trouble.

The wall may not be thick enough to house two deep boxes back to back, however. In that case, just offset the new box up, down, or sideways from the original.

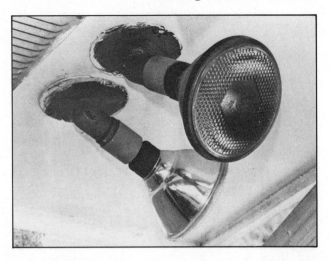

Floodlights under the eaves make great yard lights and are not difficult to install. Use weatherproof boxes and weatherproof sockets as shown here. Find a "hot" line in the attic, and run a cable from it down through a wall to a box for a switch to control the lights. Connect the black wires to the switch, splice the white wires together, and continue the cable from the switch box to the light socket.

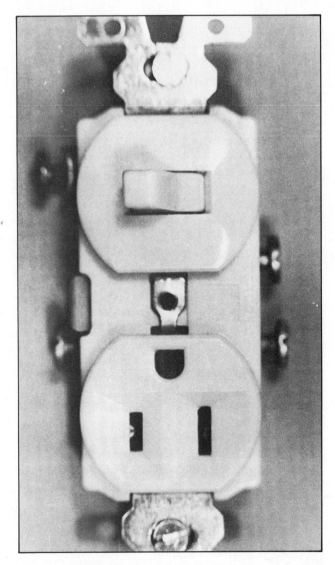

If you have a lighting circuit with the "hot" wire going to the switch box, this combination switch and receptacle is an easy way to add a receptacle. Note that one side has a silver colored screw and a brass screw. The other side has two brass screws with a metal bar between them. Connect the black hot wire to one of the two brass screws with the metal bar between them. Connect the black wire which goes out to the light to the brass screw on the other side of the switch. The neutral white wire from the hot cable goes to the silver colored screw and on to the light. You now have a receptacle that is hot all the time while the switch still controls the light.

If for some reason you want your end of the run receptacle (only one cable in this box) to be switched, just hook the white wire to the silver colored screw on this combination switch and receptacle, and connect the black wire to the brass screw on the same side of the device as the silver screw. Nothing at all connects to the two brass screws on the opposite side. You will now have a receptacle that is hot only when the switch is on.

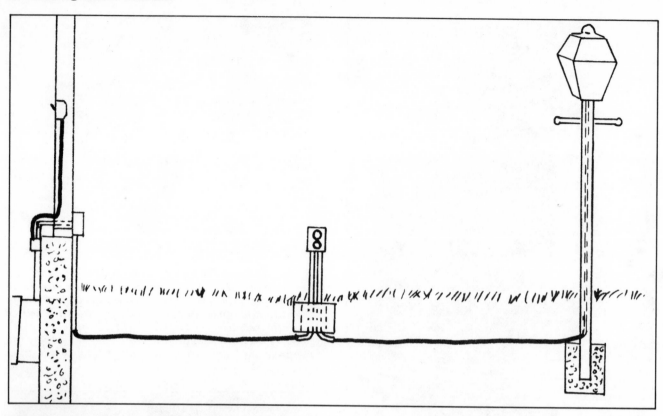

Underground wiring permits you to install receptacles anywhere outside for such conveniences as stake lights, electric lawn mowers, trimmers, bug killers, radios and TVs. Use type UF cable and run it in a trench at least 24 inches (610 mm) deep to protect it from ordinary digging. The circuit must be fused separately before leaving the house. The easiest way is to begin it at an unused circuit breaker. If none exists, install a small sub panel as described earlier.

Run the cable up the wall in conduit (if the breaker panel is in the basement) to a junction box. Run a switch leg up through the wall from the junction box if you want to control the receptacle and a post light from inside. If you want the light switched, but not the receptacle or receptacles, it will be necessary to run a pair of cables. Bring the cable through the wall in conduit to an outside weatherproof junction box and from the box into the ground in conduit. Conduit will not be necessary for any part of the run underground, except when bringing the cable back up to a weatherproof box for the outside receptacle. Here the conduit's function is to keep moisture out of the receptacle and to hold the box. Bringing bent conduit up through the hole in half of a concrete block helps keep the assembly rigid.

An outside receptacle can be added by running a cable from the nearest inside receptacle. Brick is slightly harder to work with than wood construction but by no means impossible. Remove mortar around the end of one brick, and chisel off enough of that brick to permit a weatherproof outside box to be installed. Mortar around the box to seal it after the wiring is finished. Always keep the caps on when not in use so moisture does not get into the receptacle.

The hole to accept the box was chiseled into this old wall back in the 1940s when plaster covered the bricks. The hole will be used in the rewiring, but the owner wants exposed brick walls, so he will cover the box, hole, and wire with a length of narrow pine plank.

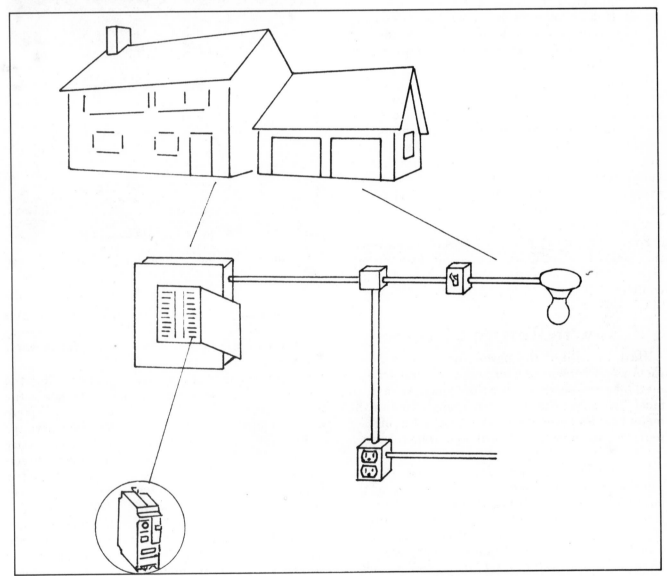

Garage wiring must be protected from overload before it leaves the house, just as other outside wiring must. The National Electrical Code requires a ground fault interrupter in garage wiring, and one place for it is in the circuit breaker panel. Then the entire circuit is protected instead of just one receptacle.

Garage circuits vary widely, depending on individual needs and desires. The minimum is one switch-controlled overhead light and a receptacle for operating tools, battery charger, etc. You may want to switch the light on either from the garage or inside the house. If so, see the section on new construction for instructions on how to hook up 3-way switches. You also may want more receptacles if you work a great deal in your garage. No special materials are required. If the garage is separate from the house you may want to run the circuit underground with UF type cable. If the garage is attached, the same materials used in your house should be acceptable in the garage.

Rewiring

Rewiring is usually done because the old wiring looks dangerous or the electrical system is inadequate. Inadequate service is dangerous, because invariably it is overloaded by a maze of extension cords connecting too many devices to too few receptacles. Houses have burned down because electric heaters were connected to already overloaded circuits by means of thin extension cords not designed to carry that much current.

Back in the planning stage, you checked the insulation of the old wire. If it was found satisfactory you probably intend to leave it as is. It will continue to supply such things as overhead lighting and receptacles for floor and table lamps. These items do not require large amounts of current. In this case, your job is not so much rewiring as the addition of extra circuits on a large scale. You also will have the advantage of continuous electrical service while you install the new circuits.

Even if the entire house requires complete rewiring, you may want to let all or some of the circuits stay intact and operating until the new wiring is completed. If you are living in the house, the need is obvious.

How to Rewire Circuits

Start by calling the power company to notify them you are rewiring your home and that you will need larger capacity service than you now have. In most instances, the power company supplies the meter base and provides you with an outline of their requirements and special local regulations concerning the service entrance.

Older homes were supplied with two-wire 120 volt 30-amp service. If three wires are coming to the house from the transformer on the pole, the service is both 120 and 240 volts (or 110/220 or 115/230) and is capable of delivering at least 60 amps. Today even 60 amps may be inadequate. You probably will be required to install at least 100 ampere service.

If 100 amperes are adequate, buy a 100 amp circuit breaker panel. However, if you can afford the higher priced 200 ampere panel, by all means select that one. Inform the power company of your electrical plans and requirements, including your need for 200-amp service.

Start with the new circuits. Wiring old circuits into the new panel will be the last step. This last step will include the lighting circuit and also any ade-

The chart on the inside of the door will tell you what each circuit breaker controls, if the electrician filled it out.

quate individual circuits (perhaps installed after the original wiring), each of which provides separate service to a clothes dryer, range, furnace, air conditioner, etc.

Some reasons for inadequacy were probably the lack of the two required 20 amp kitchen circuits, the 20 amp laundry circuit, possibly enough adequate size receptacles in the family room, and perhaps individual circuits for furnace and air conditioner motors. Or maybe you are providing adequate service so you can add central air conditioning, electric heating, or an electric hot water heater. Whatever your plans, run those new circuits now.

Each circuit begins at a circuit breaker. The special 20 amp circuits to kitchen and laundry, other 20 amp number 12 wire circuits to supply additional outlets, a separate circuit to a furnace motor or room air conditioner, etc., each start at a single 20 amp circuit breaker.

Circuit breakers are devices that open a circuit when too much current flows through it. Circuit breakers are simply reset when problems are solved. They do not have to be replaced with fuses. Breakers are available in many sizes but primarily 15 amp and 20 amp are used for 120 volt residential circuits. These are single pole breakers that only switch on or off a single conductor (black wire). For 240 volt service two pole breakers must be used. This turns on or off both ungrounded conductors. They may be 15, 20, 30, 50 amps or more.

Except where armored cable is required, 2 wire, number 12 non-metallic sheathed cable with ground will be used. The black wire will connect to the circuit breaker screw. The white wire connects under one of the screws on the common or neutral bus bar. The bare grounding wire connects under the grounding bus bar, if provided. If not, the bare wires are connected to the neutral bus bar. In the event there are not enough screws on the neutral bus, twist the bare wire in one cable with the bare wire of the next cable, and so on until all of the bare wires are spliced together. Finally, connect an end of one of these connected wires under a neutral bus screw.

A connector holds the cable where it passes through the knockout in the panel cabinet. The same type of connector will hold the other end of the cable where it enters the outlet box. Staple the cable in place within 12 inches (304.8 mm) of each box or cabinet and at least every 4½ feet (1371.6 mm) in a straight run. The cable also is stapled at each bend, corner or change in direction. Obviously, it will not be possible to staple the cable where it has been fished through walls.

When running cable the length of a rafter or a joist, do not staple it to the bottom edge where it is exposed to possible damage. Staple the cable along the side of the rafter or joist. When crossing perpendicular to the rafters or joists, again do not staple the cable to the bottom edge. It is permissible to first nail a 1 x 4-inch (19 x 89 mm) board across the rafters and joists and then staple the cable onto this board. But even this method leaves the cable exposed. It is better to drill a hole in each joist or rafter (at about its middle) and pull the cable through.

Staple the cable just enough to hold it in place, but do not hammer down the staples so hard that the insulation may be crimped or cut.

Staple the cable within 12 inches of each box. At least 8 inches of free conductor must be left in each box for making connections.

Each circuit begins at a circuit breaker. Single pole breakers such as these supply 120 volts. Connect the black wire under the screw on the breaker. These are the "hot" wires. If number 12 wire is being used, choose 20 amp circuit breakers. Wherever number 14 wire is selected for the circuit use only 15 amp breakers.

The white wire in each cable is the "neutral" wire. It is always connected to the bus bar as shown here. The bus bar is grounded for safety by connecting it to a water pipe or a rod driven into the ground. It is also connected directly to the breaker panel cabinet which grounds the cabinet for safety.

Staple a run of cable at least every 4½ feet (1371.6 mm). When running the length of a rafter or joist, staple the cable on the side of the rafter or joist, not on the bottom edge where the wires might be subject to damage.

When running cable across rafters or joists, drill holes to accept the wire, as shown. It also is permissible to install a 1 x 4-inch (19 x 89 mm) board across the bottom edges of the rafters or joists to use as a raceway for stapling the cable. But do not simply staple cable across the board's bottom edges, as the wire becomes too vulnerable to damage.

All number 12 wire circuits which are designed to handle up to 2400 watts (and this includes individual circuits for furnace motors, room air conditioners, etc.) are protected by 20 amp circuit breakers. If all existing circuits in the panel are at capacity or if there is no room for additional circuits, another panel will have to be added. Please refer to the chapter on adding extra circuits.

Other size circuit breakers will be required in special circuits for ranges, stovetops, built-in ovens and clothes dryers. All of these items operate on 240 volts, and this means 3-wire cable.

Ranges need 50 amp circuit breakers. Run 3-wire number 6 cable unless otherwise specified by the manufacturer or local codes.

Stovetops and built-in ovens require 3-wire number 6 cable if both are connected to the same circuit.

In this case, the combination will need a 50 amp circuit breaker protection. Often, however, it is much easier to run separate circuits to each. In this case, each circuit may be a 3-wire number 10 cable protected by a 30 amp double breaker. Some large ovens might require 50 amp double breakers. Check the manufacturer's specifications.

Clothes dryers, if they are 240 volt units, also must have 3-wire number 10 circuits protected by 30 amp double breakers. For 120 volt units, check the manufacturer's specs.

Hot water heater requirements vary from 2-wire number 12 cable for small 120 volt units to 3-wire number 10 cable for bigger 240 volt units. Check the manufacturer's specs.

Surface Wiring

Occasionally it is impossible to hide the wires. Unfinished attics and basements are good examples. Sometimes it is extremely difficult to get wires through the walls of older homes. And sometimes, it does not even pay to try. That is what surface wiring was designed for.

In areas such as basements and attics where appearance may not be especially important, ordinary non-metallic sheathed cable is used in conjunction with either surface boxes or non-metallic surface receptacles, switches, and light sockets.

Surface boxes are different from ordinary outlet boxes in that the corners are rounded so things won't catch on sharp corners. The boxes house ordinary switches and receptacles. Special covers with rounded corners also should be used to avoid corners.

The non-metallic surface devices are quite different. They are in themselves the complete unit—box and device—and need nothing more but to be held to a surface by screws and have wires attached. Although they are not as attractive as boxes with decorative covers, they are less obtrusive than boxes.

All circuits requiring 240 volts need double circuit breakers. The breaker panel is arranged so when a single breaker is snapped into place, it automatically connects to 240 volts. The 240 volts are present between the two screws of the double breaker. If you are using 3-wire cable that has red, black, and white wires, connect the red and black under the breaker screws (it does not matter which wire connects under which screw) and the white to the neutral bus bar.

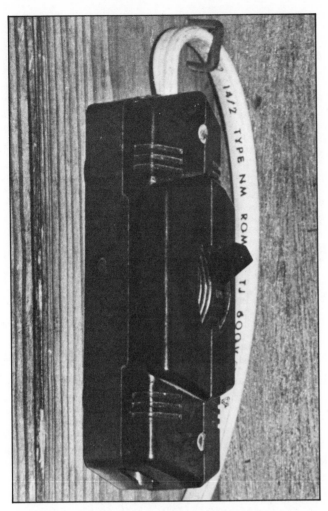

Surface switches are easily installed in surface wiring situations and are less in the way than most metal boxes.

This is an end of the run surface receptacle. The cable enters from the right. But note the other holes in the left end. These are for extra cables that enable us to connect up to eight or 10 of these receptacles in parallel in a single circuit.

Surface wiring is done where wires cannot be hidden and in areas where appearances are not important. This non-metallic socket takes the place of both box and socket.

They blend in better with the surroundings and almost vanish from notice until needed and looked for. They can be easily and quickly installed if permitted by local codes.

Another surface wiring solution to the problem of getting through walls is the raceway. Two types are available: metal and non-metallic.

The metal raceway serves the same purpose as conduit. It holds and protects wires. But the raceway is rectangular and more attractive than the round conduit. It is also easier to install. One type comes in two pieces: a channel that is fastened to the wall or baseboard, and a cap that is snapped into place after the wires are laid into the channel. Special fittings are available for making corners.

Metal raceways can be installed on, or on top of, baseboards. One type can be used in place of a baseboard: it positions the outlets the required distance from the floor, according to code specifications. Metal raceways also may be used in basements as they provide grounding safety. A flat beveled-edge raceway carries wires over floors. It resembles door sills in shape. Non-metallic raceways are also avail-

able. Check local codes to be certain these surface wiring solutions are acceptable.

In practice, raceway circuits usually begin with the wires already in an outlet box in the wall. Some raceways even plug into an existing receptacle instead of being wired directly. If old wiring is used, be sure not to overload the circuit.

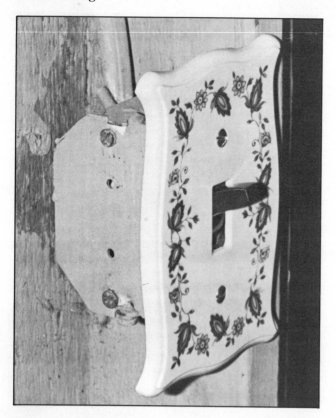

Sometimes it is necessary to use a box instead of a surface device in order to use an appropriate decorative cover. In this instance, an ordinary deep box placed the switch at the best position with maximum mechanical strength. In places where the boxes are more exposed, it may be better to select special shallow surface boxes with rounded edges.

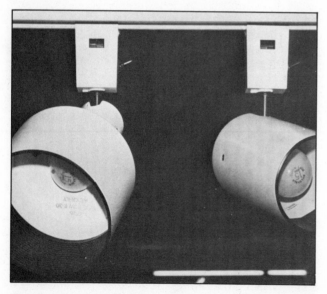

Track lighting is a system designed for store displays, but it has been adopted by some homeowners. The track may begin at an existing box or a new box and go across a ceiling or wherever desired. It is essentially one continuous, long receptacle. Specially designed light fixtures can be plugged and snapped in place in one operation.

Adding an outside light is not much different than adding an inside lighting circuit. Locate a nearby inside receptacle. Run the cable from it up to an opening in the wall for the switch. Attach the black wires to the switch, and pass the cable through a hole cut through the wall and siding. The main difference is the box to hold the outside fixture. It should be a special weathertight box made for this purpose.

Connecting Old Wiring to a New Panel

When the rewiring job is completed and the new circuits are wired, all that remains is to hook up the old circuits that you intend to keep.

Begin the job of connecting the old circuits by disconnecting the main fuse or breaker in the old panel box. All circuits are now dead. Disconnect all circuits from the old panel except the hot wires from the meter. Do not touch these wires until the power company disconnects the old meter.

All circuits except the incoming hot wires are considered branch circuits. Pull these branch circuit wires out of the old panel and run them to junction boxes. Use two or more boxes if necessary to hold the splices. Splice new cable to the old inside of the junction boxes. If there are enough empty breaker spaces in the new panel, simply run each spliced cable to a new breaker. If the old wiring was number 14, use 15 amp circuit breakers.

When there are not enough circuit breakers in the new panel, plan to use the old fuse or breaker panel as a sort of substation distribution box. In this case, do not begin work until the power company has removed the old meter to disconnect the power from your old panel.

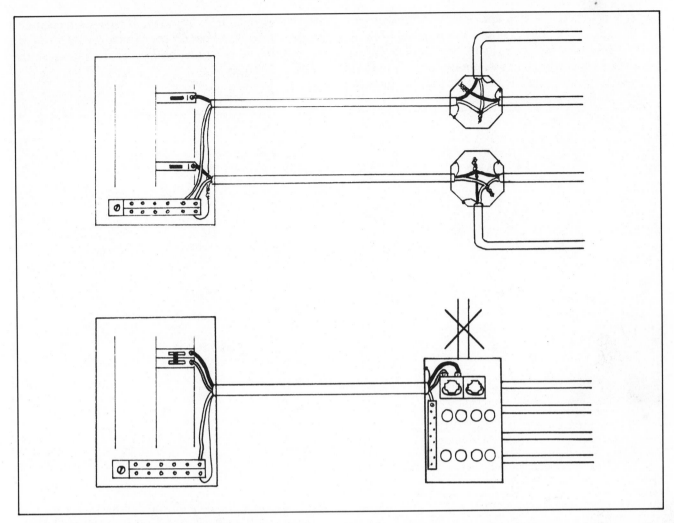

Sometimes old wiring is not bad, just overloaded. We may want to use portions of the old circuits for lighting while using new circuits to supply electricity to devices that require more power. In this case, just keep those parts of the old circuit intact which supply electricity for ceiling, floor, and table lamps. These circuits may be disconnected from the old fuse box and spliced together in junction boxes as shown in the top drawing. Run cables from these junctions to circuit breakers in the new panel. If the old wiring is number 14, use a 15 amp breaker for each new circuit. The number of old circuits which can be joined in a single junction box is determined both by physical capability of the box to hold the wires, plus how many lights will be operated on such a newly joined circuit. When number 14 wire is used, stay well below the 1800 watt total limit for all the lights on a single joined circuit.

It is also permissible to leave the old fuse box intact and simply disconnect the wires which came from the meter, as shown in the bottom drawing. In place of those wires, run an appropriate size cable (number 6 if the old service was 60 amps) to an appropriate size breaker (50 amp if the old service was 60 amps) in the new panel. The National Electrical Code requires that if this method is chosen, the neutral bus bar be unscrewed and insulated from the cabinet.

All old branch circuits will remain hooked up to their original fuse or breaker connections. Remove the old incoming wires from the now disconnected meter. If the service was 240 volt, 60 amp, replace the incoming wires with 3-wire number 6 cable. Make the connections and run this new wire to a 50 amp double breaker in the new panel. If the original service was only 240 volts, 30 amps, it will be suitable to use 3-wire number 10 and run it to a 30 amp double breaker in the new panel.

The main problem for the amateur in using the old panel is that the National Electrical Code requires that such sub panels must not have their neutral bus bars connected to their cabinets. The sub panel cabinet must be grounded back to the main panel by means of the bare grounding wire like any other junction or outlet, but the metal bus and cabinet must not be connected together. The neutral bus is connected to your cabinet. It should be removed and insulated from the cabinet. This new problem may send you back to the new panel looking for enough breakers to go around.

In the junction boxes (mentioned in the first system) you can splice two old circuits into one new cable going to a single breaker in the new panel. But if you do this, keep in mind that you are not to increase the breaker size above what it was for one old circuit. If the old circuits were number 14 wire, and you connect two circuits to one breaker, the old breaker is still only 15 amps, not doubled. These old circuits cannot be used for anything but lighting and must not be overloaded. If there is a chance of receptacles being used other than for lighting, disconnect them or put a solid cover over them so they cannot be used.

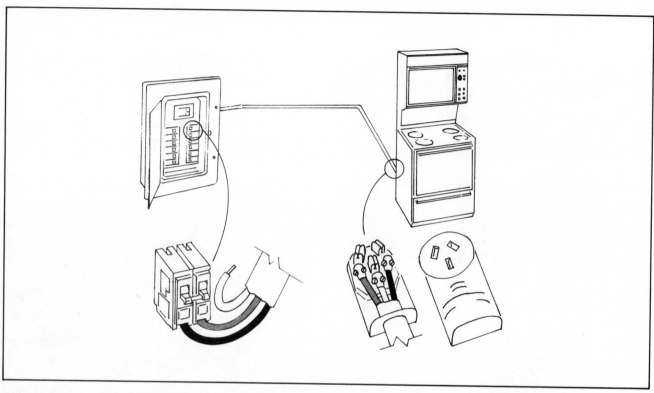

Rewiring may include the addition of an electric range. Begin a range circuit at a double (240 volt) breaker. Use 3-wire number 6 cable. Connect the white wire to the neutral bus bar, red and black to either screw on the breaker. Install a 50 amp receptacle with the white wire connected in the middle as shown.

If separate stovetop and built-in ovens are installed, it may be easier and cheaper to run a different circuit to each. Begin at 30 amp double (240 volt) breakers, use number 10 wire and install 30 amp receptacles.

The Service Entrance

3-Wire 120/240 Volts

Modern installations require 100 amp minimum. Most electrical devices in the home use either 120 volts or 240 volts. To make sure that all these devices receive the proper amount of electricity, three wire, 120/240 volt wiring is necessary.

Refer to the illustration showing the three battery hook-ups. Circuit "a" is one flashlight battery with wires connected. Between those wires we can measure 1.5 volts. Circuit "b" shows two flashlight batteries hooked in series. The voltage across the two wires now adds up to 3. Combine the two circuits as in "c." Across the two outside wires, we can still measure 3 volts. And from the center wire to either outside wire we can get 1.5 volts. We now have a 3 volt system for wherever that is needed, plus two 1.5 volt branches. With three wires, it takes the place of the six wires that would be required if the two 1.5 volt circuits and the 3 volt circuit were separate.

But, battery circuits are direct current. In our home, we are dealing with alternating current. But you get the idea. The illustration shows a transformer out on the pole that has 7000 volts coming to it from the power company.

First look at "a." The primary coil has 7000 volts applied. The two secondaries each furnish 120 volts. To get 240 volts, connect the two secondaries together as in "b." Not much different than the battery circuit, is it?

Under the right conditions, 240 volts can blast you right out of your socks. Ground is neutral—zero volts. If you are making a good connection to ground—say standing barefoot in a puddle—and you grab 240 volts, it is "good-bye." You get the full 240 volts and with it comes all the current it can shove through your body.

Power is brought from the transformer to the house through the weatherhead.

The wires run through the weatherhead down to the meter box and into the house.

To minimize this potential shock hazard, we need to connect part of the secondary circuit to ground to make it zero volts. This is done by grounding the center wire. Our house circuits will still be delivered 240 and 120 volts, but with greater safety.

By grounding the center wire, we mean precisely that. We make a physical connection to ground. In the city, that center wire will be connected to a metal water pipe that comes out of the ground into the home. If the meter is between this connection and the pipe going into the ground, the meter will have to be bypassed with a heavy wire to insure a good electrical path to ground.

The metal water pipe must extend at least 10 feet (3 meters) out into the earth to make an acceptable ground connection. This grounding system may not be possible in rural areas where wells supply the water through plastic pipe. In this case, it is necessary to drive a ground rod 8 to 10 feet (2.4 to 3 meters) into the earth to make a good connection. Check with the power utility for their grounding requirements, as well as any local code requirements.

The white neutral wire is connected to ground. The National Electrical Code permits a bare wire to come from the transformer to the meter and from there to the ground rod or water pipe. Check local codes, however. It may not be approved for aluminum wire. It may be permissible to run a bare neutral wire into your circuit breaker panel from the meter.

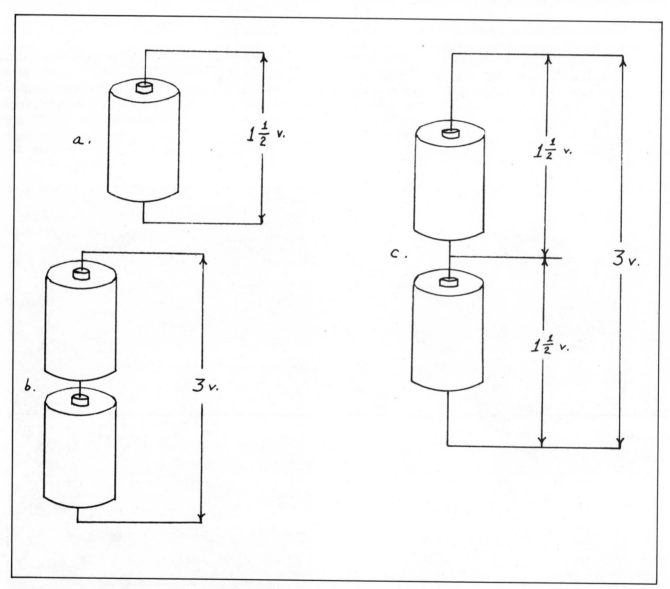

To learn how the 3-wire 120/240 volt system works, examine these three simple circuits. The first is just one battery that supplies 1.5 volts. We will compare that to a 2-wire 120 volt circuit. In "b" we have two batteries connected in series which doubles our voltage to 3. If we doubled 120 volts, we would have 240 volts, of course.

How do we get both 1.5 volts and 3 volts (as we get 120 volts and 240 volts in our homes) and do it with only three wires? By simply hooking up a center wire between the batteries, as in "c." We now have a 3 volt source and two separate sources of 1.5 volts.

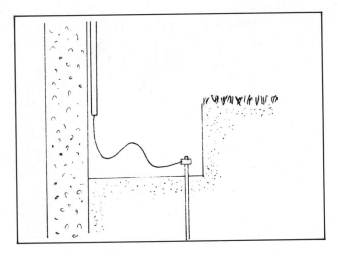

In rural areas where water is pumped from wells and the pipes are often plastic, it is necessary to drive a rod 10 feet (3 meters) into the earth to get a good ground connection. Whenever running any kind of wire underground from a building, it is a good idea to allow slack and shape the wire into a rough "S." The weight of dirt fill or the strain of freezing and thawing will not break such wires. The trench should be dug about 2 feet (.6 meters) deep before driving the rod to ensure a good connection deep in the ground.

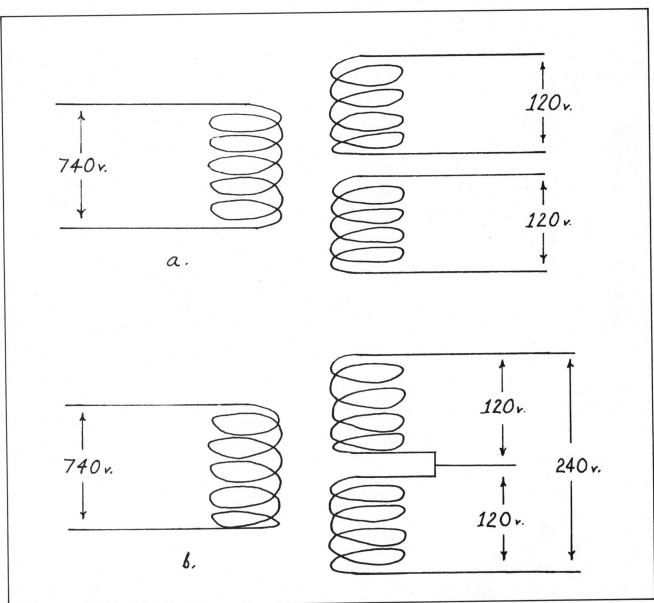

For a more accurate explanation of how the 3-wire 120/240 volt system works, compare the battery circuits in a previous illustration to these two circuit diagrams of transformers. There are 7000 volts across the "primary" coils. The "secondary" coils (a) have the right number of turns of wire to develop 120 volts in each. If we connect the two secondaries in series (b) we develop a source of 240 volts, plus two separate sources of 120 volts.

The center wire of a 240-volt system is grounded for safety. In the city, this is usually done by making a connection from the neutral bus bar in the circuit breaker box to the nearest water pipe. If the metal water pipe goes at least 10 feet (3 meters) into the earth, it makes a good ground. When a meter is between the connection and where the pipe enters the earth, it must be bypassed with a jumper wire as shown here to assure a good ground.

Learn to read your meter. Watch it. You will realize how much it turns daily, and you will more likely make a conscious effort to conserve energy.

Service Entrance
Location and Installation

Where the two ungrounded conductors from the power company transformer and the grounded conductor are brought into the home is the service entrance. The service entrance is where the power utility makes its connection with your main fuse or circuit breaker panel. The grounded conductor is never connected to the main circuit breaker but to the neutral bus bar.

With this basic understanding, the next step is to call the power utility to request service for your home. They will probably help you decide the best location for the service entrance. Possibilities can be limited by location of the power lines and physical obstructions. It is not desirable, and sometimes not permitted, to run the wires over the roof to the opposite side of the house. And neither you nor the power company wants extra poles in the yard to carry service around the house.

When it is decided which corner of the house will receive the service entrance, determine where to locate the circuit breaker panel. The breaker panel must be as close as possible to the meter. The big wires from the meter base into your breaker panel are expensive and difficult to work with. The number 10, 6, or whatever wires running to ranges, water heaters, clothes dryers, etc., are also more expensive than the number 12 circuits. These should be as short as possible.

Once you and the power company agree to a meter location, they will give you a meter base. In some areas you purchase the meter base. They will also supply you with a drawing that illustrates an acceptable service entrance installation. It is generally up to you to supply all materials except the meter base.

Ask the power company about any special local codes you need to know. In some rural areas, you may be on your own. The power company simply has you sign a release that says they are not responsible for damage caused by your wiring mistakes. In general, the more your area is populated, the more regulations you will encounter.

If you are building a new home and are requesting electricity for the building site, the power company will set up temporary service. A meter, a small breaker box, and a weatherproof outlet will be installed on an outside pole. You will get your power for construction work from long, heavy-duty extension cords connected to the outlets.

In most areas, you will need a building permit before any construction, including electrical work, begins. Ask the power company if a permit is necessary and who to contact.

You will have two choices regarding how the power company will bring power across your prop-

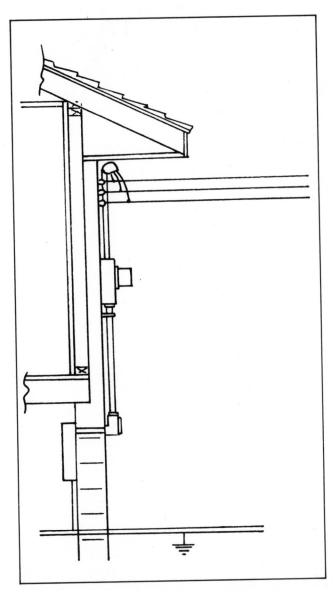

For new construction, call the power company, and they will install a meter with temporary service on an outside pole. You will have power to operate tools from long, heavy-duty extension cords.

When you contact the power company for service, they will probably supply you with a drawing illustrating how they want you to install the meter base and other materials associated with the service entrance. This drawing shows a city installation which is grounded to a water pipe. In this case, overhead wires were selected.

Overhead wires may not be aesthetically pleasing. In many areas, they are not necessary.

erty to the meter: overhead or underground. In some very rocky and rural areas, you may not have the underground option. Underground wires are not as dangerous since they do not fall during wind and ice storms.

There are disadvantages to underground service, mostly to the power company. Heat dissipation is not as rapid underground as in the air, so power losses in the line to your meter are greater. And if a tiny hole opens in the insulation, the wire is burned in two.

Better insulated wire is required underground, and service repairs are more expensive. While these are power company expenses, all costs are eventually passed along to the customers.

Unless the area is very rocky, you may be able to have the service entrance wires run underground to your house.

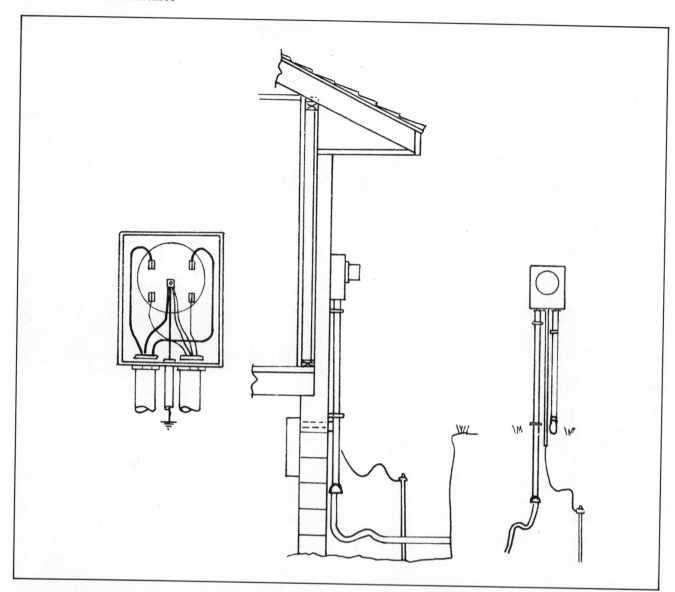

This drawing is the type the power company may give you for an underground service entrance. In this case, we have shown a ground rod as used in most rural installations. Of course, a water pipe ground also could be used with an underground service entrance. And a ground rod could be substituted for the water pipe ground in the overhead installation in the previous illustration.

Overhead service is brought in through a weatherhead. The power company lines in this case are attached to insulators fastened on the mast. Because they are below the weatherhead, the wires must go up. Rain cannot run down into the weatherhead. If water ran into the weatherhead, it could cause a short circuit in the meter base.

Always install the meter box at eye level so the meter can be easily read.

Power company wires are always attached to the top connections in the meter base. The wires already connected in this picture are those going into the circuit breaker panel inside the house. Cables coming from the large conduit at left are connected to the transformer. Note that wire with white insulation was not available, so a black wire was marked with white adhesive tape to indicate it is the neutral wire. As you can see, the neutral wire is attached directly to the meter base by means of a connector bolted to the metal cabinet. The ground wire also is fastened to this connector.

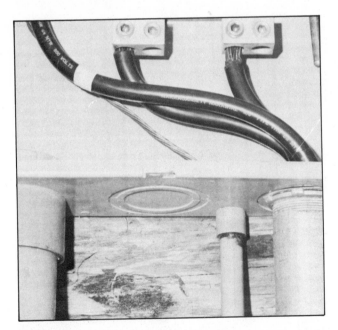

These power company lines are supported by a large, bare "messenger" wire, which in this case is simply tied around the mast. More often, a connector is used to attach the messenger wire to the mast in a more solid, enduring manner.

In this close-up it is easy to see how the bare ground wire runs down through the small conduit which extends into the earth where the wire is attached to a ground rod. Also note how the large wires involved in a service entrance are held by screws in heavy-duty connectors.

Service Entrance Wires

The service entrance wires are fastened to the top set of connectors inside the meter base. Wires connected to the bottom will run through another length of conduit and into the house by the shortest route to the circuit breaker panel.

These wires are large, stiff, and difficult to bend and pull around corners, so "ells" are used to make it easier to make the corner from the conduit into the house. The ell is a 90 degree fitting with a removable cover. This allows more working room to push and pull the wire to get it through. The ell must have a weatherproof cover. Use only one ell to get through the wall directly to the breaker box. After bending wires through one, you will not be anxious to try another.

We have not specified the size of these large service entrance wires because that depends upon the ampere capacity of the service entrance. Your power company will tell you what size and type of neutral, ground, and hot wires to install in your particular service entrance. Conduit size also varies according to wire size.

It may not be possible to buy red, black, and white wires. If only black is available, tape one to show it is being used as the white wire. It (the white wire) must always be the neutral and must be connected to the center ground lug in the meter base. The other two wires, whether red and black or both black, are connected to the right and left lugs. It doesn't matter which wire is connected to which side. The system will work either way.

The same thing applies when you get the wires pulled inside the circuit breaker cabinet. Whether the red or black wire hooks to the right or left side doesn't matter. But the white wire, or the wire marked as a white wire, must connect under the large screw provided for that purpose on the neutral bus bar. All other white wires coming from all the branch circuits in the house will also connect under screws in this bus bar. Unless a separate bus bar is provided for the bare grounding wires from the various circuits, they will also connect to this neutral bus bar.

If you are rewiring a house and still have service through the original meter, there is no need to get

The "ell" is a right angle fitting with a removable weathertight cover. The cover is removed so the heavy wires can be brought out, looped and then shoved through the conduit in the wall.

Inside the circuit breaker panel, the two hot wires go to the main circuit breaker. Current goes through the main breaker (when it is in the "on" position) to the two metal bus bars below it.

power turned on to the new meter until the entire job is finished. The same is true in new construction where temporary service already has been provided.

Installing the branch circuits is covered in other sections. All that remains is to connect these circuits to their respective circuit breakers. Punch out the necessary "knockout" holes in the breaker panel. Slide connectors on the cables so the sharp edges of the holes in the cabinet will not cut through the insulation. Run the wire inside the cabinet, allowing plenty of length to reach the correct slot on the breaker panel, and fasten down the connector. Connect the white and bare wires as described earlier, and connect the black wires under the screws of the proper circuit breaker.

If the local building code calls for an inspection, the inspector usually wants to see the job at this point. The wires are run to all outlet boxes, all branch circuit wires are in the breaker panel, but no breakers, no switches, no receptacles, no fixtures or anything is connected.

Simply pushing a circuit breaker to the "on" position may not activate it. Instead, push down to the "off" position as far as it will go, and then flip it to the "on" position.

To install the breaker in the panel, notice that one end has a hook that fits under a hook strip in the panel. Slide that end in place first. Then push down on the other end of the breaker until its metal connector snaps in place.

This illustration shows the two bus bars directly below the main circuit breaker. There are 240 volts between these bars and 120 volts between each bar and the neutral bus bar. Between the two hot bus bars you can see "blade" connectors insulated by black plastic. Each of these blades is connected underneath the plastic to either the right or left bus bar. As can be seen, two breakers can be connected to each blade, each one providing 120 volts to a circuit. The hooks punched out of the metal on the far left and right are insulated from the bus bars and are used merely to help hold breakers in place.

Breakers are installed by sliding one end under the hook and snapping the connector end onto the blade.

Circuit breakers are devices to protect circuits when excessive current flows. Unlike a fuse which blows and must be replaced, breakers trip to an off position and can be reset. Single circuit breakers are for 120-volt circuits, double for 240-volt circuits.

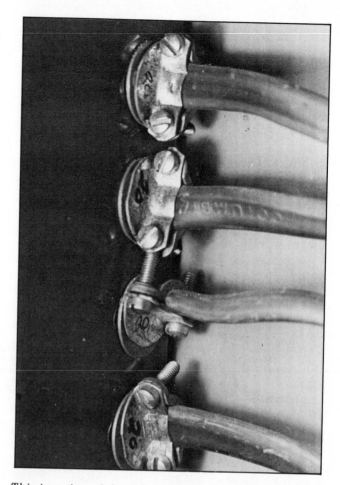

The 120-volt branch circuits have two insulated wires, plus bare grounding wires. The black wire of each connects to a circuit breaker. The white wires all connect to the neutral bus bar. All bare grounding wires also connect to the neutral bus bar unless there is a special grounding bus bar provided. Note the connectors which hold the cables as they pass through the circuit breaker cabinet.

This is a view of the cable-holding connectors from outside the cabinet.

New Construction and Hook-ups

Wiring a house in the process of being constructed is not much different than rewiring an older home. It is usually easier. Everything is exposed and easily accessible. Cables do not have to be fished through walls. Just lay the cable in place, staple it to studs and joists, and the job is done.

Since the wiring is exposed throughout the work, it is easier to trace circuits, and understand how the system should work. The simple circuits previously discussed also apply in new construction. The circuits used in new construction and described here are also used in rewiring and adding new circuits.

Outlet Boxes

There are two types of boxes: outlet and junction. But junction boxes may be used as outlet boxes. To avoid making a very simple thing seem complicated, choose either the rectangular outlet box or the octagonal junction box (it almost looks square, but with the corners shaved off). This octagonal box may also be used with a hanger between ceiling joists to hold a fixture. Or it may hold a simple porcelain socket type fixture in the basement, attic, or closet. Where you are installing boxes for switches and receptacles, choose the rectangular boxes.

There are several varieties of both boxes. Some have built-in clamps to hold the cable so movement which could cut insulation is avoided. Others require connectors. Some rectangular boxes are made to be nailed to a surface. You may want those in an unfinished basement or attic. Others have brackets which make it easier to fasten them to studs. And others are simply mounted by sliding two nails through holes in the sides and driving these nails into the stud. Hangers are available for mounting boxes between studs or joists.

Some boxes are shallow; others are deep. Choose the deepest box that will fit. (If boxes must be back to back in a wall, for example, there is not room for two deep boxes.) The deeper the box, the easier it is to squeeze the necessary wire into it without crowding the receptacle or switch.

If you have a special need, ask an electrical supplier. Special boxes are made shallow, deep, extra strong for chandeliers, big for multiple devices, narrow for tight places, and for just about any purpose imaginable.

Consult the blueprints or your own plans and begin fastening boxes in place. Receptacle boxes should be about 12 inches (304.8 mm) above the floor. Switch boxes are placed about 48 inches (1219.2 mm) above the floor.

Mount the boxes so their front edges will be flush with the plaster or plasterboard which will be in-

Do not install switches, sockets, and receptacles while the wiring is being done. The inspector will want to see the "rough" wiring before any devices are installed. Fasten rectangular outlet boxes in place, run the cables, and allow at least 8 inches of wire to extend from each box.

The octagonal (or square) box may be used either as a junction box where wires are joined or spliced, or as a box to hold a ceiling fixture.

This octagonal box is being used as a junction box. Junction boxes are necessary as places to join wiring where three or more cables come together. Ordinarily, that many wires would overcrowd the average switch or receptacle box. This particular junction box has built-in cable clamps.

stalled after the rough wiring is completed. Adjustable mounting ears come on most boxes to compensate for varying thicknesses of plaster. They also can be reversed. In one position they mount on top of the wallboard. In the other, they are fastened against lath, allowing room for the thickness of the plaster between the ears and the front of the box.

In most instances, the wire used in boxes will be non-metallic sheathed cable. It must be stapled within 12 inches (304.8 mm) of each box and at least every 4½ feet (1.3 meters) during the run. This is not required when cable has been "fished" through finished walls. When running cable perpendicular to joists or studs, drill holes and pull the wire through. On parallel runs, staple the cable to the side of the studs or joists as discussed earlier.

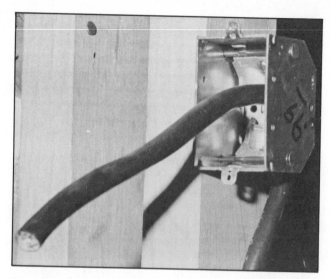

This box is simply fastened to the studding with two large nails.

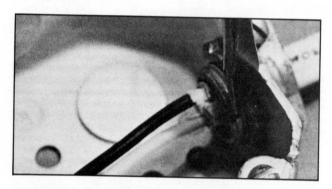

Tighten the connector nut with a screwdriver so the cable is held firmly in place.

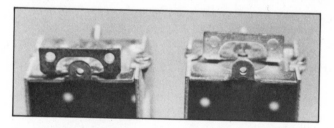

Adaptable mounting ears come on most boxes. They either reverse or slide back and forth. In one position, they are flush with the plasterboard surface. In the other position, they can be fastened against the lath, with room allowed for the thickness of the plaster.

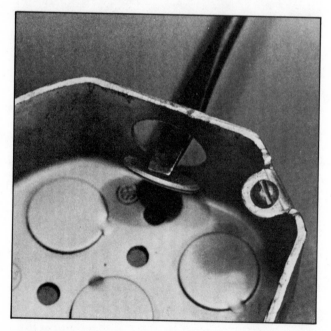

This type of junction box does not have built-in cable clamps. Knock-outs are provided in strategic locations on the box so holes can be opened easily to admit cables. Connectors are required to hold the cable as it enters so the insulation is not cut by sharp edges.

Twist-outs are easily removed with a screwdriver to permit the entry of cables into switch and receptacle boxes.

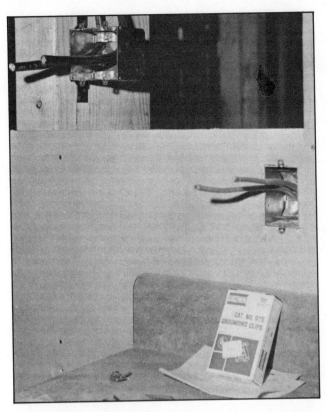

Boxes can be ganged easily by removing a side wall from each and then joining them together. Loosening one sidewall screw makes that wall removable. Shove the boxes together, tighten the screws from each box, and a double box is formed for use where double switches or receptacles are needed.

A variety of switch and receptacle boxes are available. The box, above left, is held to the studding by brackets and small nails.

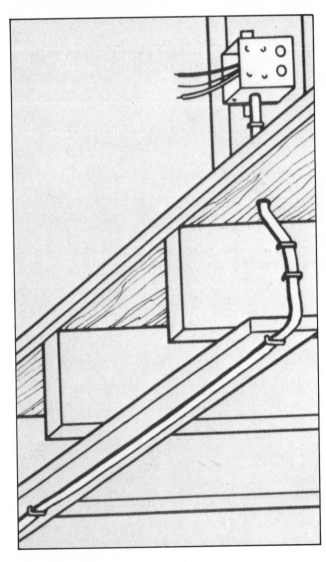

In most wiring situations, use nonmetallic, sheathed cable. Run it through or along the sides of joists and rafters, not along or across the bottom edges. Staple this cable at least every 4½ feet (1.3 meters) and within 12 inches (304.8 mm) of each box.

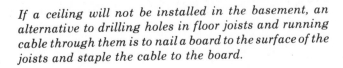

If a ceiling will not be installed in the basement, an alternative to drilling holes in floor joists and running cable through them is to nail a board to the surface of the joists and staple the cable to the board.

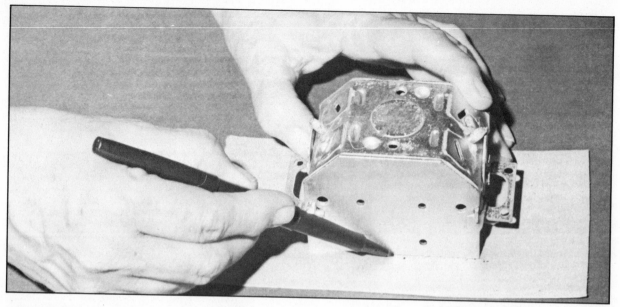

Make a template of the boxes you are using by tracing one of them on a piece of cardboard. Cut it out, and use it as a pattern to mark the cut-out for boxes on plasterboard.

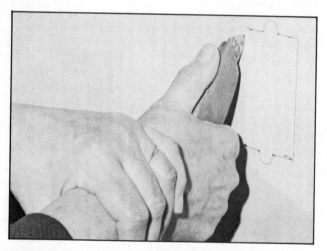

A utility knife can be used to make the cuts in plasterboard. Grip the wrist as shown for added control and force.

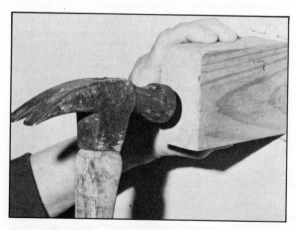

After the cuts have been made, a hammer and a suitable size block of wood can tap free the cut-out section.

A sabre saw also can be used to cut openings for boxes in plasterboard or plywood. Drill two or more holes large enough to accept the saw blade, and cut along the lines.

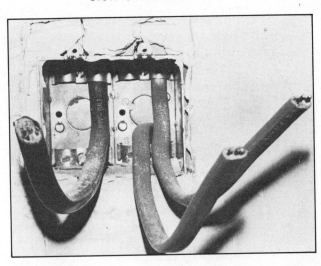

If holes must be cut larger than the boxes, or if it is done by accident or error in measurement, patching plaster may be used as fill. Stuff a little steel wool into the hole first (keep it out of the box or it may short the wires) and then add plaster patch. The steel wool will give the patch support to prevent cracking from weakness, as happened here.

If openings in plasterboard are accurately cut to the right size, there will be no need to fill in holes. The switch or receptacle plate will cover it.

Using Conduit

If conduit is required or desired, you will probably choose thinwall. It is easily bent by a conduit bender, is not difficult to cut with a hacksaw, and uses threadless pressure-type connectors and couplers. Rigid conduit, on the other hand, must be threaded like water pipe which makes it more difficult and time-consuming to work with.

Bend and install the conduit first. Before pulling wire through, remove any rough edges or burrs from all conduit to prevent conductor insulation from becoming damaged. Bending conduit with the wires inside could damage the insulation. No splices are allowed within the conduit. Splicing must be done in junction boxes.

Thinwall conduit must be supported with special straps within 3 feet (.9 meters) of each box and at least every 10 feet (3 meters) during the run.

Bending conduit is not difficult, but do not attempt it without a regular conduit bender. Properly used, the bender will not kink the conduit. Other methods of bending almost always will.

The bender has a hook on one end of the arc to grab the conduit. The opposite end of the arc has a place to put your foot to hold the bender tightly against the conduit while the lever is pulled toward you to make the bend. Some benders have built-in levels to indicate when the 90 degree bend is complete.

If you have tried bending conduit without reading any instructions, you have already learned there is a trick in making the measurements come out right. Suppose you need to make an 18 inch (457.2 mm) drop from the basement ceiling to an outlet for the washer. You will notice an arrow on one side of the bender. Do not place that arrow on the 18 inch (457.2 mm) mark. If you are using ½-inch (12.7 mm) conduit, subtract 5 inches (127 mm). Place the arrow on the 13-inch (330.2 mm) mark. When the 90 degree bend is made, the end of the conduit will be 18 inches (457.2 mm) from the portion of conduit your foot is standing on. The bend has raised the conduit by 5 inches (127 mm).

When using ¾-inch (19.05 mm) conduit, figure 6 inches (152.4 mm) in the bend, and for 1-inch (25.4 mm) conduit, 8 inches (203.2 mm). Use as few bends as possible in the conduit. Bends increase the difficulty of pulling wire through conduit.

When strapping conduit on the face of concrete, use one of the straps as a guide to drill holes with a masonry bit. Fit the holes with plastic expandable inserts. These will hold the screws which fasten the clamp and conduit against the concrete wall.

Most residential type conduit runs are short and pose no problem in getting the wire through. Sometimes it can be easily pushed through. Long runs will have to be pulled through with a fish tape. If there are many bends (the Code allows up to four 90 degree bends in a run unbroken by boxes), it may be necessary to add a junction box to serve as a "pull box" or opening from which the wire may be pulled through the conduit.

Bending thinwall conduit is easy, but do not attempt it without a tool made for that purpose. Properly used, the bender makes a neat elbow with proper radius for ease in pulling wires through. Without a bender, you will kink the conduit.

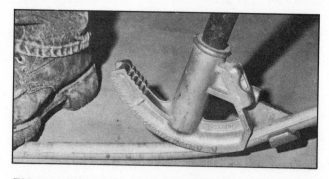

The electrician here has kept his foot out of the way so you can see how the bender works. In practice, it is best to place your foot on the knurled section at the rear of the bender to hold the assembly firmly against the floor. This helps prevent kinking.

Note the bubble of the type ordinarily found in carpenter's levels. When the bubble is centered as shown, a perfect 90 degree bend has been formed.

Thinwall conduit is much easier to work with than the rigid kind and is chosen where conduit is required in most home wiring. It can be cut with a hacksaw or with a pipe cutter as shown here. Tighten the cutter a small amount each time it is turned around the conduit, and before long it will cut through the pipe. Burrs inside the conduit, caused by cutting, should be reamed out before installing the conduit. If not done, these sharp edges can cut through the insulation when the cable is pulled across them.

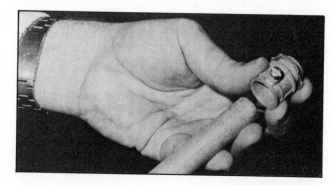

Slide a connector onto the end of the conduit, and tighten the set screw. Shove the threaded end through a hole in a box and use the lock nut to hold it in place. The assembly can be seen in the photograph at the left.

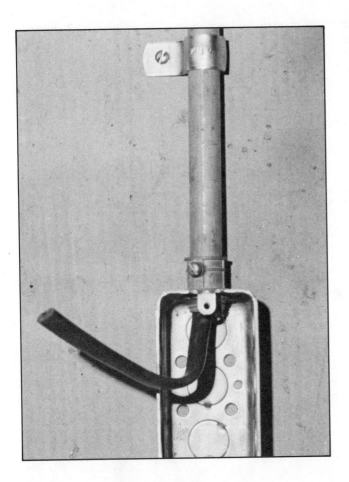

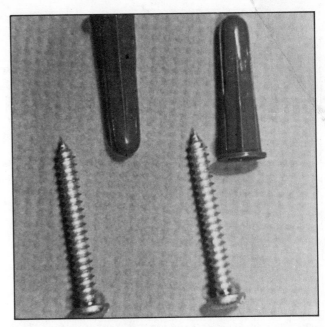

Plastic expandable inserts are placed in the holes in the concrete. The screws expand these inserts tightly against the walls of the holes, holding the clamps and conduit in place with great strength.

When strapping conduit to the face of a concrete wall, use one of the straps as a guide to drill holes in the right place with a masonry bit.

Hook-up How-to

In earlier sections, you learned a simple battery and bulb circuit and its comparison with a light and a switch. Later you learned that the voltage from a source (now our breaker panel instead of a battery) could be run around the house and tapped into at several places. A group of receptacles is such a circuit. All they do, actually, is act as connectors to join other devices (lamps, irons, TVs, etc.) with the current flowing through the cable. Receptacles are mechanical devices to tap into voltage-carrying wires without making splices. Start with one of these circuits. They are simple, easy, and do much for the beginner's confidence when he learns that something he did actually works.

Two receptacle hook-ups are illustrated. A cable comes from the breaker panel to the first receptacle. Another receptacle is added on. This can go on 8 or 10 times, if it isn't a special circuit for the kitchen or laundry.

It isn't always convenient to begin at the breaker and keep adding receptacles. Sometimes it is easier to bring the source cable (the one from the breaker supplying the voltage source) into the middle of the outlets and branch off to either side using a junction box to make the split. Electrically, everything is hooked in parallel. White wires connect to white wires, and black to black.

Suppose we have seven receptacle outlets installed. There is no handy place for an eighth. But we do need an overhead light on a switch in this room. This is accomplished by hooking the cable to the last receptacle (or perhaps from a junction box if it is handier to make a split) and run it to the switch box. Another cable from the switch box to the light ends the circuit. Hook it up as illustrated.

Note that the black wire, the hot wire, is broken by the switch. The white wire, the neutral, splices right to another white wire and goes on to the light. Never is the hot wire to go to the light first and the switch to be in the neutral leg of the circuit.

If the hot wire is hooked to the socket and you touch the metal socket while changing the bulb, you may receive a severe shock even if the switch is turned off.

If it is more convenient to run from the last receptacle to the light fixture first and then to the switch, that is OK, but it must be wired in the special way as illustrated.

The same thing can be done with one or two receptacles on a switch. Most receptacles are not switched, of course, but you may want one or more switched so you can turn on a table lamp or two as you enter the room. Outlet boxes can be ganged (fastened together) so there is no problem having two switches side by side for both overhead lighting and table lamps.

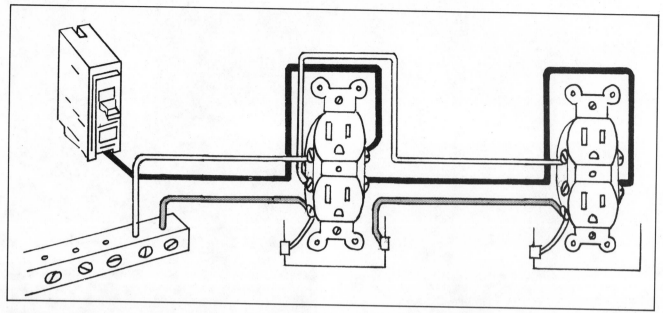

Every circuit begins at the circuit breaker or fuse. The hot wire is black. The white neutral wire and the bare ground wire connect to the neutral bus bar. For clarity, we are showing the hook-up here as if there is no outer sheath on the cable. And we have drawn only a part of the box for grounding at each receptacle to keep the drawing simple and uncluttered. We have shown the hook-up of only two receptacles, but a total of eight or ten are allowed on a single circuit. Simply add the others just like the second receptacle is wired to the first. Grounding wires connect from box to box and from box to the green screw on each receptacle. White wires connect under the silver colored screws, black under the black screws.

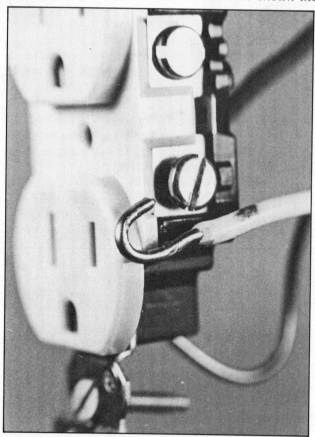

When connecting wires under screws always bend the wire clockwise in the direction the screws will turn. This receptacle is constructed so that wires can be connected either under these screws or by inserting them into holes on the back side.

This 3-way switch is hooked up by inserting the bare ends of the wires into holes. Under the holes are tension connectors that grip and make adequate electrical contact.

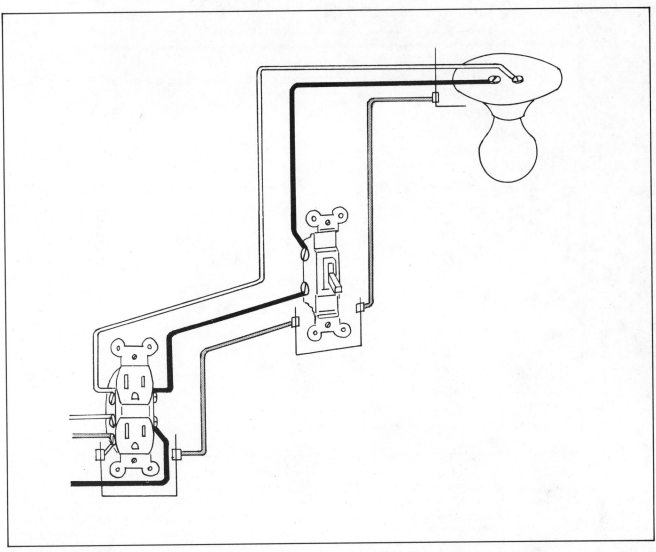

This is the end of a circuit that began at the breaker panel and ran to several receptacles. Now, however, we need an overhead light, so we continue the circuit from the last receptacle as shown. Again, grounding wires are included, and boxes are indicated by lines. Be sure to hook up the colors of the wires as we have here.

Sometimes it is necessary to continue a circuit from a receptacle in two different directions. But that would be too many wires crowding a receptacle box. Use a larger junction box to make the splices. Again, the drawing includes pictures of how to hook up the grounding wires to both boxes and receptacles. We have drawn the complete junction box, but the receptacle boxes are represented by lines.

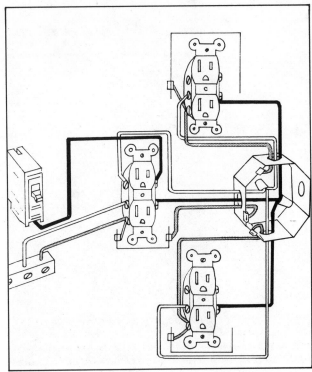

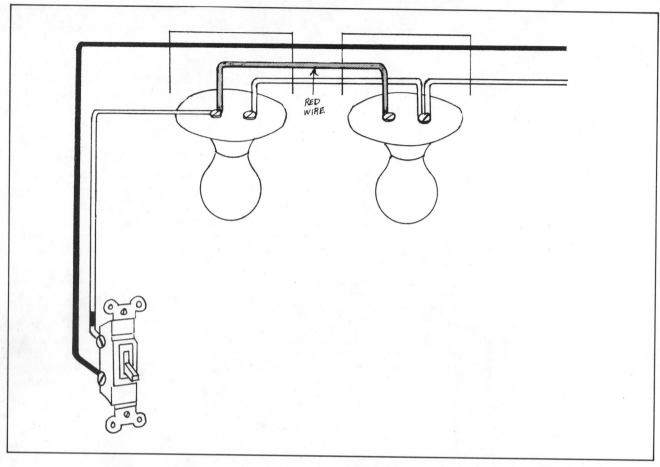

RED
WIRE

If it is more convenient to bring the hot line from the circuit to the light than to the switch, do so. But hook it up with the white wire (neutral) going to the light and the black to the switch. You will end up with the necessity of connecting the white wire in the cable to the switch. Just wrap it with a turn or two of black tape to indicate that you are using white as a hot wire whenever the switch is turned on.

In this drawing we have shown how to use a length of 3-wire cable in order that two lights can be wired in parallel and controlled by a single switch. If only one light was needed, the red wire would be unnecessary. The white wire from the source or circuit would connect to the light socket. The white wire in the cable between the light and switch would connect to the other side of the light socket and to the switch. For simplicity, we have not included the grounding wires. Hook these wires as indicated in previous drawings.

If you hooked up the wires the wrong way, you could have this basement light connected with the power at all times instead of only when the switch is turned on. You could accidentally touch the metal socket while screwing in a bulb and get a severe shock.

Grounding wires can be attached under one of the screws in the box or under a grounding clip, as shown here.

3-Way Switching

The next circuit—3-way switching—is guaranteed to mystify if you just look at the switches and try to imagine what they are doing. So, let us look inside the switches. They may be spring loaded for positive switching, or they may make silent contact with mercury. But despite physical differences, inside, electrically, they are just common, easy-to-understand single-pole, double throw switches.

Look at the first of the circuits in the 3-way hook-up illustrations. The single poles on the switches can be thrown to close with one contact or the other. Trace the current path from the hot wire through the switches back to the neutral wire. Obviously, the switches are on and the light is burning. Now use your imagination. Mentally flip switch A to the other side. You cannot trace a complete circuit, so the light is off. But with switch A still flipped in that off position, flip switch B. Now you can trace a complete circuit again. At all times, you can turn the light on or off from either switch without having to touch the other.

Again, however, it is not always convenient to wire a circuit in a certain way. It may be easier to run the hot line into the light outlet first or run the hot wire to a switch box and then place the light outlet between the switches. As illustrated, 3-way switches can be hooked up in several different ways.

Since 3-wire cable is necessary in these circuits, you will probably find red, black, and white wires. And once again, white wires will sometimes end up on switch legs and black wires on fixtures. There are cases where it cannot be avoided. And the National Electrical Code permits it in the switched circuits. However, always make sure the white neutral wire from the breaker panel goes to the light. In one hook-up method, the white wire will go directly to the light. In the others, the white wire must travel

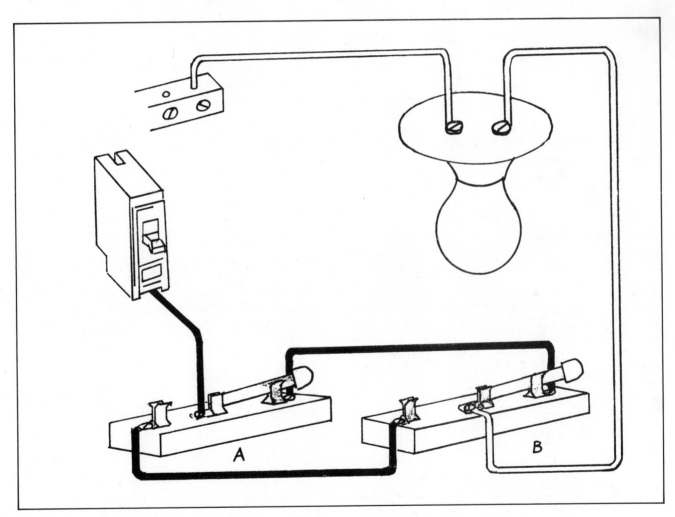

To make 3-way switches understandable, we see they are no more than single-pole, double-throw switches. Trace the circuit with your finger from the circuit breaker to the "common" pole on switch A. The knife blade is thrown to the right, so follow the electrical path through it to switch B and on through the light and back to the neutral bus bar. As you can see, we have a complete circuit, so the light is "on."

Mentally, throw one of the switches to the opposite side. Try to find a complete circuit. There is none, so the light is "off."

Now throw the other switch to the opposite side, also. You can trace a complete circuit, so the light is on again.

from one box to the next before it arrives at the light. The second rule dictates the hot black wire from the breaker panel always goes to the nearest switch. When connecting the black wire, it goes to the so-called "common" connection on the switch which means we are connecting this black wire to what amounts to the same thing as the blade in our sim- ple single-pole, double-throw switch. This common connection is not at the same place on all 3-way switches, but instructions in this regard will come with the switches. The screw head on this connection also will be a different color than the other two. It will be black or copper while the other two are brass or silver-colored.

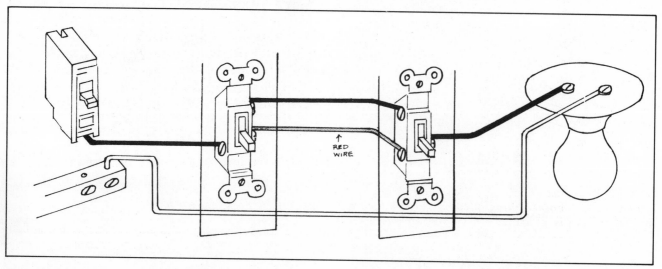

This is the same 3-way switch hook-up as shown in the previous illustration. However, we are using conventional 3-way switches. In this case, for easy identification, we will consider the common connections to be the single screws by themselves on one side of the switches. In practice, this will not always be the case. But one screw will be different. One may be copper while the other two are brass. The copper, in that case, is the common connection.

As you can see, a 2-wire cable is necessary from the breaker panel (or continuation of that circuit) to the front switch. A 3-wire cable is needed between the switch boxes, and only a 2-wire is needed again out to the light. The third wire is invariably red. Hook them up as shown, being sure to always continue the white wire from box to box to the lamp without being broken by a switch.

You will be using a grounding wire if required by local code. We have omitted it here to make the 3-way connections easier to visualize.

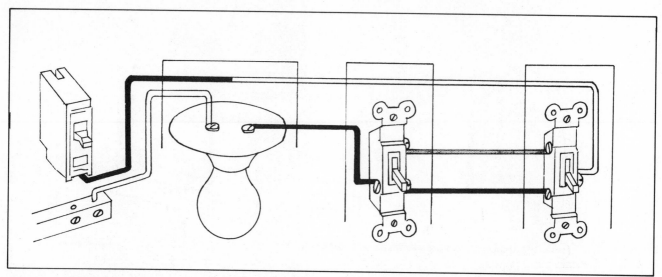

It is not always convenient to run the hot line directly to the first switch as shown in the previous drawing. In this illustration, it is handy to run the hot line or cable directly to the light. As always, the white wire from the neutral bus bar connects to the light and the "hot" black wire to the switch. The National Electrical Code requires the other wire on the light be a color other than white.

The reason is so you can easily identify which of the two wires in the light fixture box really is the neutral. For that reason, the black wire in the cable going to the first switch must be connected to the light socket. It also means we must splice the white wire to the black wire in that box.

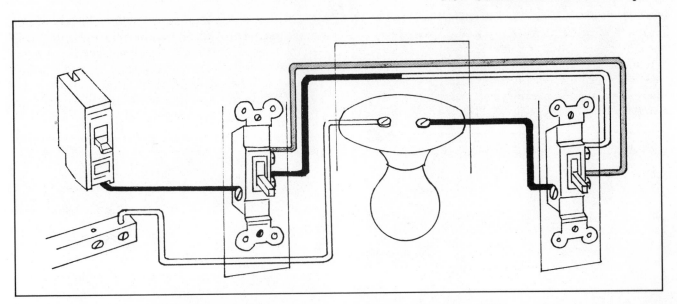

In the previous 3-way hook-ups, it was handy to have the light at one or the other end of the circuit. In this illustration, the light is physically positioned between the switches. Hook up as shown, splicing black to white again in the light fixture box.

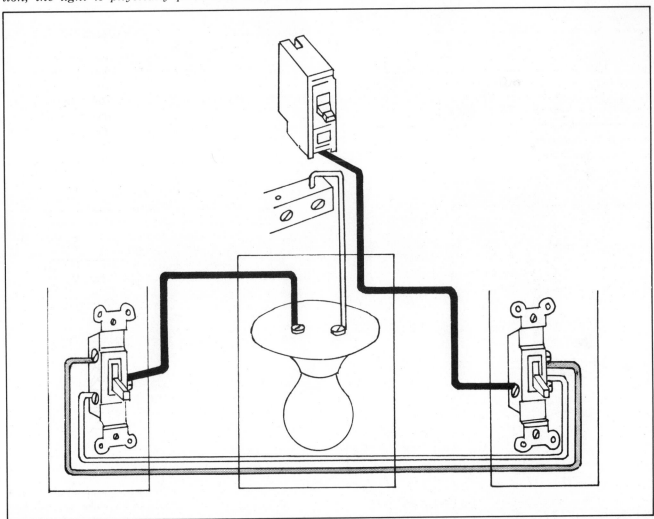

In our final illustration on the methods a 3-way hook-up can be done, we again find the light fixture box between the switch boxes. But this time it is most convenient to run the hot line to the light box instead of to a switch. One 2-wire cable comes in to the light fixture box, 3-wire cables go from it to each of the switches. The splices in the boxes are easy to remember: white to white, red to red, and black going to the commons on both switches.

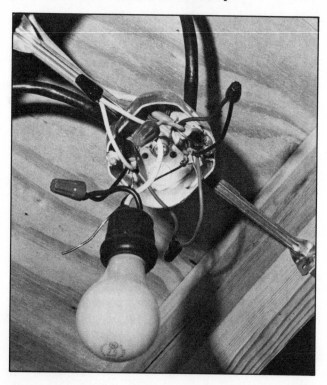

This is how the connection shown in the previous drawing actually looks in the light fixture box. A pigtail socket is temporarily hooked up here to see if the circuit works and to provide light while other construction is being completed.

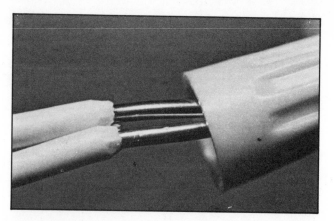

The easiest way to join wires is to use wire caps. Hold the two wires parallel.

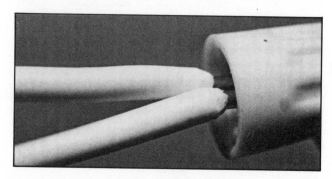

Slip on the cap and twist.

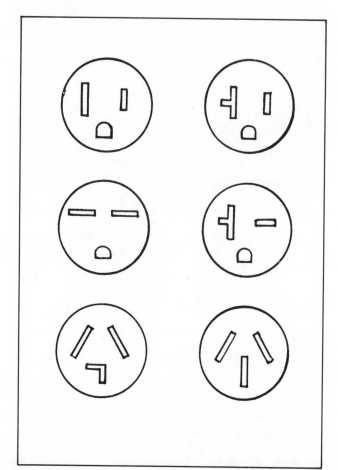

This illustration shows the various configurations of the receptacles commonly used in home wiring.

Upper left: 15 amp, 125 volt with grounding, standard for light loads.

Upper right: 20 amp, 125 volt with grounding, used for large tools and air conditioners. Smaller appliances with plugs designed for the 15 amp socket can also be plugged into this receptacle. But plugs designed for this 20 amp receptacle will not fit in the 15 amp socket because of the extra protrusion on the wide blade of the plug.

Middle left: 15 amp, 250 volt with grounding, used for 240 volt air conditioners and heavy-duty tools using 240 volt motors.

Middle right: 20 amp, 250 volt with grounding, used for air conditioners, tools, etc., having motors requiring 20 amp sockets for safety.

Bottom left: 30 amp, 125/250 volt, 3-pole, 3-wire, normally used for clothes dryers in homes.

Bottom right: 50 amp, 125/250 volt, 3-pole, 3-wire, used for electric ranges.

Dimmer Switch

An energy conserving switch (single pole or 3-way) is the dimmer switch. It provides total light variability from off to bright for as much as 600 watts. A cheaper 300 watt unit switches to two positions of light intensity. The energy savings relates to the brightness setting of the unit.

Bulbs may last longer in dimmer circuits because they operate much of the time at lower voltages and therefore, lower temperatures.

Dimmer switches are marvelous devices for changing the light level to suit the mood or need. They are easy to install and fit ordinary switch boxes. The single-pole switch has just two wires to connect instead of the two screw connections on ordinary switches. The hook-up, of course, breaks the black hot wire from the breaker panel as with all other switches.

There is a 3-way dimmer switch, too, but only one of these is used in the circuit with one other ordinary 3-way switch. The light can be turned off or on from two locations, but dimmed from only one.

A dimmer switch is available for fluorescent lights, but rather extensive conversion is necessary. The ordinary ballast in the fixture must be replaced with a special dimming ballast. And these are available only for 40 watt, 48-inch (1219.2 mm) units. In addition, the lamps must be replaced with 40-watt rapid-start types, unless of course the fixture being converted is already a rapid-start unit. The fixture must be grounded, too. Given the limitations on units that can be converted, and the costs, time, and effort involved, this has not become a challenge that do-it-yourselfers are anxious to try.

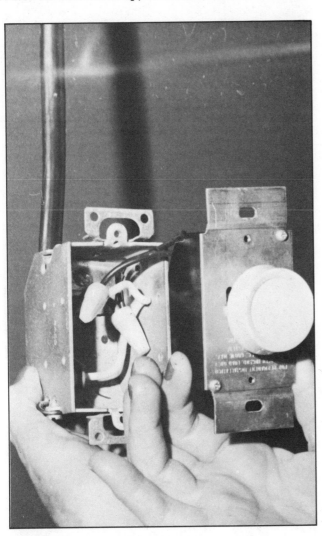

Dimmer switches have wires instead of screws or tension connections like ordinary switches. Simply connect them with wire caps, or splice and wrap with tape. Both wires on the dimmer switch probably will be black, and it makes no difference which one is connected to which of the wires in the cable.

For limited wall space, compact dimmer switches are available that are half the depth of conventional dimmers for fast, easy, installation. Thick-film microcircuitry has reduced the dimmer's depth to only 25/32" inside the outlet box. Photo courtesy of Leviton Manufacturing Co., Inc.

A combination of lighting is best. This kitchen has overhead incandescent lighting above major work areas, fluorescent lighting under the wall cabinets to illuminate countertop work areas, and a chandelier above the dining area. Photo courtesy of the Tile Council of America

Possible Lighting Arrangements

On these and the following three pages several finished rooms are presented to show just a few of the lighting arrangements that are possible by following the general electrical wiring principles and procedures described previously. The options are limitless and up to you.

Small indirect lights positioned strategically around this kitchen-dining area highlight various work areas and artifacts. Photo courtesy of Armstrong Cork Co.

The overhead lighting in this living room is provided by several incandescent lamps, recessed in the ceiling, and mirrored ceiling panels. Photo courtesy of Armstrong Cork Co.

The fluorescent lighting in the decorative box is all the light needed to adequately illuminate this dining counter. Photo courtesy of the Tile Council of America

Energy-efficient fluorescent lights are used in this basement recreation room. Photo courtesy of Masonite Corp.

Fluorescent ceiling panels permit uniform, soft lighting throughout a room. This type of lighting is excellent in dark rooms with little or no natural light. Photo courtesy of Georgia Pacific

Track lighting, generally used in studios or living rooms, is very effective in a bathroom setting. The lights can be positioned to provide overall lighting or lighting for specific functions such as shaving or applying makeup. Photo courtesy of American Olean Tile

Track lighting is used very effectively in this attic studio. The individual lights can be adjusted to provide light wherever necessary. Photo courtesy of the Tile Council of America

The dimmer switch-controlled, adjustable track lighting in this living room is perfect for highlighting any of the room's major features such as paintings, bookshelves, or the grand piano. Photo courtesy of Masonite Corp.

Electrical Repairs

I wish it were possible to put this chapter on the cover of the book so people could realize at a glance how easy it is to save money on electrical repairs. There are millions of people who know absolutely nothing about electricity, and most of them will experience electrical breakdowns. If they bought the book only for this section, and tucked it away until needed, I could guarantee two things: savings of many times the cost of this book, and self-reliance. It is terribly annoying for either men or women to be dependent upon somebody else to do even simple tasks.

If you notice female hands doing the more simple electrical repairs in the illustrations, do not get the notion that I believe it takes a man to do the more complicated work. Electrical repairs have been traditionally left to the man. But men are not always around when needed, some are quite afraid of electricity, and changes in social traditions are making it necessary for women to be more self-reliant.

For the Chicken in All of Us

Let us call it prudence. If we know absolutely nothing about electricity, we have the good sense not to touch anything that is not turned off or unplugged. When the device is turned off or unplugged, many persons are still afraid to touch its wires. Most folks will touch any part of a lamp they have unplugged, yet when confronted with replacing a bad receptacle, they are afraid to loosen its screws even after personally switching off the circuit breaker or removing the fuses. If you have this fear, I hope you will eliminate it by reading the earlier chapters in which we discuss how electricity works.

Don't lose your caution, however. I would never advise anyone in bare feet standing on a basement floor to even open a fuse box. You could possibly touch the wrong thing and get a severe jolt. Wear shoes. With rubber soles, if possible. This insulates you from becoming a path to ground for the electrical current if somehow you touch a hot connection. If you want even more insurance against such accidental shock, stand on a dry board.

If you read earlier chapters, you know there must be a complete circuit for current to flow. If you touch a hot wire with one hand, and no part of your body touches anything that can complete the circuit, you will not be shocked. We are talking about ordinary household voltages, of course. For those who feel one hand is too many to be touching anything electrical, wear rubber gloves. They will insulate your hands against mistakes and will give you confidence to proceed with the work.

You will really have to go out of your way to get shocked while changing a fuse, though. The wires are in the box under a fastened-down cover. You are only opening a door in that cover to expose the fuses, not the connecting wires. The metal rim on the fuse is in no way touching any electrical connections. The only way to get a shock is to stick your finger into the fuse socket.

Changing Light Bulbs

There is also a chance of getting shocked while changing a bulb in a miswired socket if the metal part is touched while screwing a bulb in or out. But only if you are standing on concrete with bare beet or damp shoes with leather soles. Standing on a dry board would prevent this. So would wearing dry shoes with rubber soles. Shutting off the power would prevent it, too, but few people will do that just to change a bulb. And, of course, you will not get shocked by changing a bulb in even a miswired socket if the floor you are standing on does not provide a path to the ground. Essentially, only concrete or bare dirt offers this danger. Wood or tile does not.

Changing Fuses

Now that you know changing a fuse is safe, and how to make it doubly so, the next question is how do you determine what fuse to replace? Use a flashlight for a good, close look.

The gap in the metal of the center fuse in the illustration could mean the fuse just became exhausted from long use. Replace it and go about your business. The common, small, round fuses screw in and out just like a light bulb. If your main fuse is blown, it will be a cartridge-type fuse which will snap in place. If the fuse blows frequently, it means the circuit is nearly or slightly overloaded. Try to remove something from that circuit. Plug one of the appliances into a different circuit, and see if the frequent fuse blowing stops.

When the fuse window is cloudy, this is an indication of extreme current flow. Chances are strong that something in that circuit has shorted out. Re-

Nothing could be easier than changing a fuse. Just unscrew it as you would a light bulb and screw a new one back in. Always replace the defective fuse with the same ampere size, never a larger size.

Pulling out the main fuses or circuit breakers will disconnect the power and allow you to change a switch or receptacle without any danger of shock. There is a second fuse block that resembles the main, but it is marked range. It, of course, is the protection for your electric stove.

placing the fuse without finding the trouble will probably just waste another fuse. Instead, unplug all appliances or devices on that circuit. Now replace the fuse. If it blows again with everything unplugged, the trouble is in the house wiring. More likely, it is not. You replace the fuse and now the lights work again. To see which appliance or device is bad, check them over for something obvious—an acrid odor, bare wires in the cord, etc. If you cannot see anything wrong, plug them in one at a time. When you blow the fuse, you know which appliance was responsible.

Often, the problem turns out to be crumbling insulation on the cord of a lamp, appliance, or other device. This could also happen on a ceiling light fixture, or even on the wires to a switch or recepta-

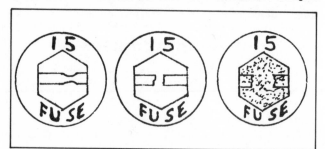

If part of your household wiring has suddenly gone dead, check the circuit fuses. The fuses which look like the one on the left in this illustration are good. The thin strip of metal is intact. In the center is an "open" fuse. The metal strip overheated and melted in the center. Chances are, you are overloading the circuit. Try to plug some of your electrical devices into a different circuit, or remember not to turn so many things on at the same time. If a fuse looks like the one on the right (the glass face is clouded) something really zapped it. Whatever you plugged in last may be shorted. If the fuse blew suddenly without anything new being plugged in, one of the devices may have just developed a short circuit.

cle. This possibility, however, is much more remote unless the wiring is extremely old or too much bare wire has been left exposed in outlet boxes.

If nothing at all appears wrong, consider that the fuse may have blown simply because too many things were turned on at the same time. Suppose it is a summer morning. The toaster is heating, the coffee maker is going, a TV is giving the morning news, and it is just a bit warm, so you plug a fan into the same circuit. A motor requires much more current to start than when it is running. If the fan had been running, and the other things had been plugged in later, maybe the fuse would not have blown. But if the fuse was close to being overloaded, and we hit it with the extra starting current of a motor, out it goes. If you use several devices that require extra current when the motors start, type S or type T fuses should be used. These fuses offer extra protection for momentary overloads when motors start.

In this instance, you know what blew the fuse because it happened when you plugged in a certain appliance. Replace the fuse and try to organize the appliances so they are not all on the same circuit.

Never replace a small fuse with a larger size, such as replacing a 15 amp with a 30 amp fuse. It may be tempting because you are tired of replacing fuses in an overloaded or nearly overloaded circuit. But do not try it. Never stick a penny under a fuse to get things going again. The fuse is your protection against fire. If you defect it with a penny, you are guaranteed trouble. The size of fuse has been chosen to protect a circuit with a given size wire from becoming overheated. A larger size fuse does not offer this protection against overheating and possible fire.

Checking Circuit Breakers

When you open the door of a circuit breaker panel, you will see what appears to be a row of switches. Only the switch handles are exposed. Circuit breakers are essentially switches—switches that kick to the "off" or "tripped" position when overloaded. All of these switch handles are in the "on" position except the one that is tripped—the one you are looking for.

Check the number of the tripped breaker and look for the corresponding circuit listed in the chart on the inside of the panel door. That tells you where this particular circuit goes. Now you can check this circuit for problems in the same manner as described for fuses.

This chart is rarely found on fuse boxes because most were installed many years ago. New installations generally utilize circuit breakers, and the practice of identifying circuits has been carried out more often in recent times.

Chances are that when you try to reset the breaker, nothing will happen. This is because not all breakers can be reset simply by pushing the handle to the "on" position. In many cases, the breaker handle will be found in the center or "trip" position. It will first have to be pushed all the way to the "off" position, and then finally to the "on" position. Now the circuit should work.

In the event that the circuit still does not receive power, even after you are certain you have properly reset the breaker and have perhaps tried a second

Circuit breakers look like switches. All will be in the "on" position except the one that has been tripped by an overload. This breaker is being reset by a 9-year old girl. There is no danger of shock. The breaker below her finger is also in the "off" or tripped position. Some breakers can be reset by simply pushing the handle to the "on" position. Others will have a separate "trip" position and will have to be pushed to the "off" position and then to the "on" before they will work again.

and third time, the breaker itself may be bad. Breakers do go bad. It is good practice to turn breakers on and off periodically to help avoid corrosion and metal fatigue.

To replace a circuit breaker, first switch off the main breaker so there is no danger of being shocked. Then remove the panel cover to expose the wiring. You will be able to see where a white wire is connected under a screw on each breaker. At that same end of the breaker, you may be able to see the part of the breaker, a sort of molded-in hook, which is under a metal projection in the panel cabinet. The opposite end of the breaker was pushed down to fasten onto an electrical contact. Pry up on that opposite end from the wire to release the breaker from the electrical contact. Once it is freed, just pull the other end of the breaker out from under the metal projection.

The breaker is now loose except for the wire. Disconnect it, and reconnect it to a new breaker with the same amperage rating as the original. Shove the hook end of the new breaker under the projection, push down on the electrical connection, and you are all set.

The handle of new breakers may be in the center or "tripped" position. If power does not return to the problem circuit when you throw the main breaker switch, you may have to reset the new breaker in the manner described earlier.

Breakers wear out when tripped too often. They can go bad for no apparent reason without ever having been tripped. To replace, pry up the end that snaps into place. Pull the breaker out from under the hook on the other end. Remove the wire. Do this with the main breaker in the "off" position.

Shorts and Opens

People who know little about electricity call most electrical troubles "shorts." A short is not a general term to cover all electrical problems. It is a specific thing that means a breakdown has provided a short-cut path for the current. You will remember that there must be a complete circuit for current to flow. The "short" diverts the current so it never reaches the appliance.

A good example is a "shorted" cord to a lamp. The insulation on the cord ages, cracks, falls off, and the wires touch—probably near the base of the lamp.

The current goes in one wire, across the shorted wires, and back out the second wire. Because current takes the easiest path, and because the shorted wires provide that path, none of the current gets to the lamp itself.

Furthermore, because the short circuit does offer little or no resistance to current flow, entirely too much current will flow. If the circuit breaker does not trip, or the fuse does not blow, something will be burned out because of the excessive current. Chances are, the wires will melt and burn apart at the place where they were touching, or shorted, together. Sparks will fly.

An open circuit is the exact opposite of a short. An open means the circuit is broken. No current can flow in that circuit at all. If the wires burned apart at the lamp as described above, and one of the wires burned completely in two, that would be an open.

A "loose connection" also is an open circuit, but open just some of the time. Sometimes a bad switch with corroded or burned contacts causes a light to flicker. When the light is flickering off, the circuit is opening. If the switch opens up completely, the light will not burn at all.

Opens occur as often as shorts. Keep in mind the difference, and it will help greatly in understanding and finding electrical troubles.

When looking for a short that blew a fuse or tripped a circuit breaker, the first thing to examine is the cord on the suspected appliance or device. Insulation cracks, falls off, the wires touch, and you may have a short circuit that blows a fuse. Discard the cord.

If a ceiling light does not come on, and the bulb is good, check the switch. With the power on, hold the test lamp probes to the two screws on the switch. If the lamp remains lit with the switch either "on" or "off," the switch is open and needs to be replaced.

Simple Testers

If you only plan simple electrical repairs you will not need sophisticated test instruments. All you need to check for the presence of voltage is a little test lamp. A small continuity tester is enough to check out shorts and opens. Both devices were described in greater detail in an earlier chapter. Buy both and have them at hand when needed. They cost very little.

Replacing Plugs

The most frequent open circuit occurs at the plugs of our electrical appliances. Any connection is susceptible to becoming an open circuit. But the plug is a point of extra stress. Watch anybody at all unplugging a device. Almost everybody just grabs the cord and pulls. The connection of the wire at the plug was not intended to take this strain. Before long, it is pulled apart and an open circuit develops.

If you suspect the plug, insert it in a receptacle, grip the cord right behind the plug, and jiggle it a bit. If the lamp (or whatever device) flickers or tries to work sometimes, replace the plug.

It may be necessary to take the device apart enough to expose the wires so the cord and plug can be tested with the continuity tester. Fasten the tester's alligator clip to one of the bare wires. Touch the probe of the tester to a prong of the plug. If the tester does not light, try the other prong. If the tester does not light either time, you have an open circuit. If it does light, move the alligator clip to the other exposed wire and repeat the procedure. If one side shows an open, chances are it will be at the plug. Although it does happen, an open seldom occurs somewhere in the middle of the cord.

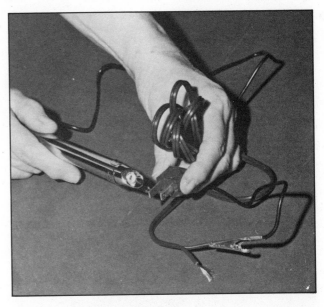

An open circuit means exactly that. The circuit is not complete. It is open somewhere. A common open occurs where the wire joins the prongs of a plug. This continuity tester is checking a cord for such an open. If it lights, the circuit is closed. If it does not light, the circuit is open. Of course, there are two wires in the cord, so make sure the end of the wire you have is connected to the prong you are checking. Or you can put both bare ends of wire in the pair of alligator clips at once. Then if the light does not come on, the wire to that prong is open.

Today, most plugs are molded right to the cord. They cannot be repaired or rewired, but they can be replaced. Just cut off the old plug with a wire cutter. Several types of replacement plugs can be attached to the cord by simply shoving the wire inside the plug and pushing down on a lever or pushing together the prongs. Metal points cut through to

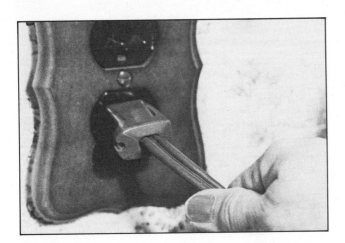

Most plugs cause trouble because they are mishandled. Many people do not grip the plug to pull it from the receptacle. They grab the wire instead, placing all the strain at the connection between the wires and the prongs.

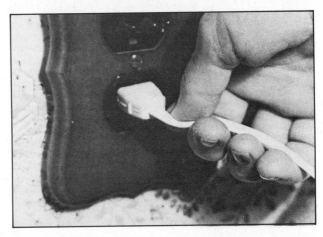

If you suspect an open or "bad" connection at a plug, insert it in a receptacle, and jiggle and push the wire around in various positions. If the lamp or whatever it is connected to flickers or tries to work, you have a plug that needs replacing.

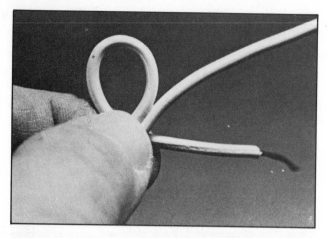

Use an Underwriter's knot if the replacement plug is large enough to hold it. Begin by looping the left hand wire and passing it behind the right hand wire.

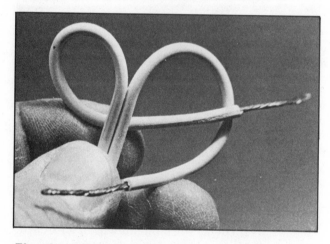

Then loop the right hand wire over the end of the left hand wire.

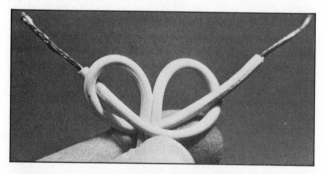

Pass the right hand wire through the left hand loop, and the knot is finished.

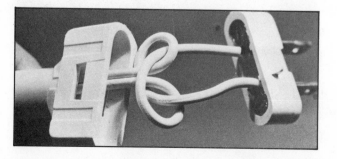

automatically make contact. These are fine. They are easy to attach. But sometimes the cord is a little odd sized, and a point fails to penetrate at the right place and it does not work. Since the point does not make good contact, the repair may not last.

The best replacement plug is one that contains screw connections that fasten down securely on the wires. Tie an "Underwriter's knot" in the cord before the connections are made. When the cord is pulled, the pressure is on the knot instead of the electrical connections.

Repairing Lamps

Plugs and cords are the most frequent trouble-makers in lamps. If that part of the circuit is in proper order, dismantle the socket. Use the tester as shown to check for open circuits in the socket. The outside metal portion of the socket is connected directly to one screw connection. The center contact in the socket is connected through the switch to the other screw connection. Connect the alligator clip to the center contact. Hold the probe to the screw contact. The tester does not light? Turn the switch. Now the tester should light. If it still does not, the switch is open. You can buy a replacement switch and socket assembly at a hardware store.

Replacing a Wall Switch

The chandelier fails to light when you flip on the switch. There are several bulbs, so you can assume all are not bad. Or maybe it is a fixture with just one bulb. You tried a new bulb. Nothing. You tested the old bulb in another lamp. It works. Everything else on the circuit is working, so you know the circuit breaker has not tripped. The likely suspect is the switch.

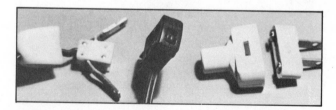

Replacement plugs come in a variety of styles. The ones at the left and middle are easiest to use. You do not even have to bare the wires. Push them in and sharp points will make contact. The type on the right makes the best, most permanent replacement.

Draw the knot up fairly tight so it will fit inside the plug, and fasten the bare ends of the wires under the two screws. Assemble the plug, and forget it. If anybody yanks it out by the cord, the knot will take the strain, and the connections will not be harmed.

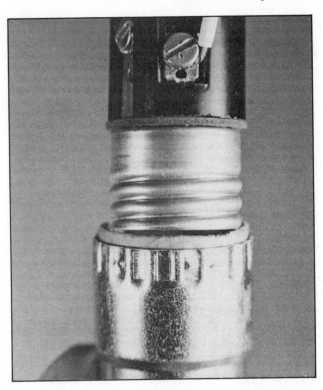

When repairing a lamp, and the cord and plug are OK, next check the socket and switch combination. Begin by prying off the end cap. Note that an Underwriter's knot has been used here to protect the connections.

Next, pull the socket free from its housing.

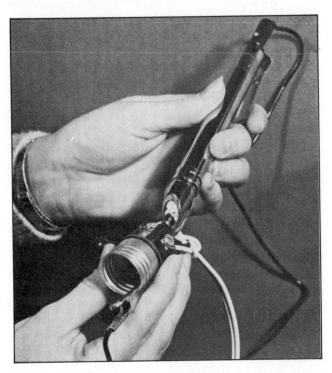

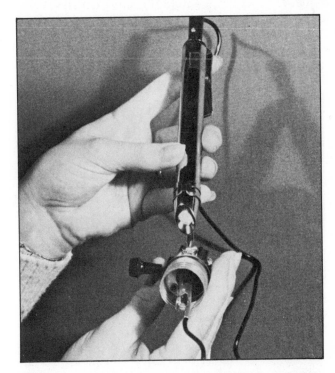

Using the test light, determine which screw is connected directly to the outside metal part of the socket. The bulb will light when you have the right screw.

The other screw makes connection to the internal switch which opens or closes the circuit to the center connections in the socket. With the test lamp connected as shown, its bulb should turn on and off when the switch in the socket is turned on and off. If the test light remains off all the time, the socket's switch is open.

Flip the circuit breaker handle to its off position or unscrew the fuse. If you are uncertain which circuit the switch is on, disconnect the main fuse or breaker. To be sure the current is off, check for voltage with the test lamp. With the switch in the "off" position, touch the probes of the test lamp against the two screws. If the switch has an extra green screw on it for grounding, turn it to its "on" position and touch one probe to the green screw and the other to both of the other screws in turn. If the little neon bulb in the test lamp never lights in any of these tests, there is no voltage present, and you can proceed.

You have already unfastened the switch from the outlet box. Now remove the wires from under the screws. Hook the alligator clip of your continuity tester to one screw and touch the probe to the other. With the switch in the "on" position, the continuity tester should light. If it does not, the switch is your trouble.

Sometimes all of this testing procedure is quite unnecessary. The switch may have been sending warning signals for some time before it quit. Maybe the light flickered occasionally or did not always come on at the exact instant the switch closed. The switch is going bad, and should be replaced.

When replacing a switch, you may want to change to a silent mercury switch. It is more expensive, but it lasts indefinitely and is silent.

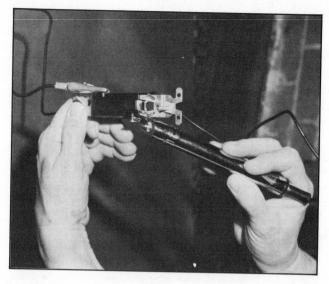

To check a wall switch with a continuity tester, shut off the power, and unhook the switch. With the tester connected to the two screws of the switch, the light should go on when the switch is in the "on" position. If it does not, the switch is open.

Replacing 3-way Switches

If one switch is showing evidence of being bad, plan to replace both switches. If one is bad, the other soon will be. Besides, not all 3-way switches look alike or have the connections in the same places. It is less confusing, now and later, to have two identical switches.

Shut off the power and unscrew the switches from their boxes but do not disconnect the wires just yet. First, look the switch over carefully. There are three screws. If there are four, the fourth is a green colored screw for the bare grounding wire. But look at the three other screws. One is always a different color than the other two, but not green. Forget about its position on the switch. Maybe one is copper while the other two are brass.

Remove the wire from that odd colored screw first. Identify that wire with a little piece of electrician's tape or white adhesive tape. Do likewise with the second switch. Now remove the wires from under the remaining screws.

Look at your new switches. Again disregard location of the screws. Pick out the one that is different, but not green. If there are directions on the box or with the switch, they will call this screw the "common" connection. Fasten the identified wires to these screws on both switches.

Now hook up the remaining two wires to each of the switches. In this case, it does not matter which wire goes to what screw.

Fasten the switches back in the outlet boxes and turn on the power. If you have followed instructions, either switch will turn the light both on and off independently.

When replacing 3-way switches, always look for the one screw that is a different color than the other two. Mark that wire before disconnecting the switch. That wire will go to the different colored screw on the new switch. It is the "common" connection. And commons are not always in the same place on different switches. The screw on the left in this photo is copper while the other two are brass. It will be more obvious than it is in this black and white photo.

Replacing Light Fixtures

Occasionally, a socket acquires a loose connection where a rivet holds one metal piece to another. Sometimes they get old and the insulation on the wires becomes dangerously brittle. But most fixtures are replaced because of a change in decoration or a need for better light.

Replacement is easy. Shut off the power and double check it with the test lamp. Do not rely on the switch to shut off the voltage to the socket. Someone may have incorrectly wired the hot side of the line to the socket.

With the power off, remove the screws that hold the fixture to the outlet box. Drop the fixture and disconnect its two wires. Simply connect the two wires from the new fixture, and mount it to the box as the other fixture had been.

If the old fixture was mounted with machine screws directly into the threaded holes in the outlet box, and did not utilize a "mounting strap," it is

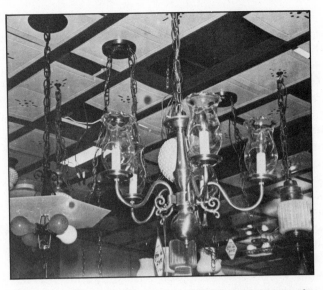

Light fixtures under 10 pounds (4.5 kilograms) may be fastened directly to the box or to a mounting strap on the box. Chandeliers over 10 pounds (4.5 kilograms) require a nipple and a hickey for mounting.

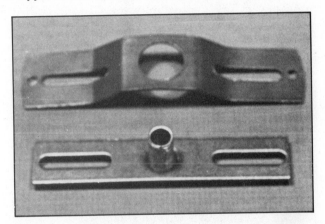

Mounting straps come in a variety of styles and sizes. A new light fixture may require a different type or size than was used with the old fixture.

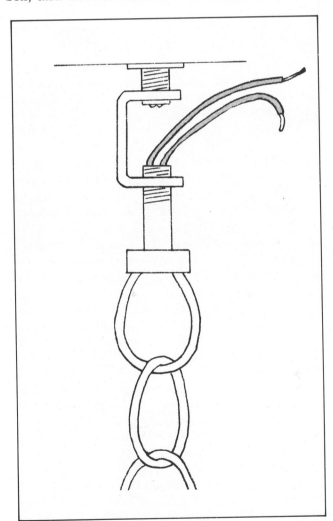

The hickey for mounting a chandelier is the C-shaped metal bracket. Threaded holes are in both legs of the C. The top screws onto the nipple in the box, and the chandelier nipple screws into the bottom.

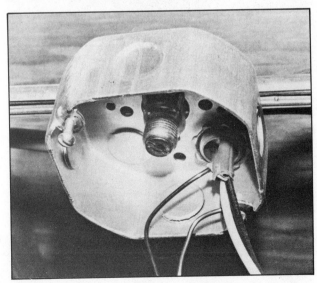

This style box has a nipple for mounting a chandelier.

possible that you will find the screw holes in the new fixture are in the wrong place. Measure the distance between them and buy a mounting strap that will fit. There will be holes to screw the mounting strap to the box. And then the new fixture can be screwed to the mounting strap.

If the new fixture has a hole in the center for mounting, you have the choice of buying a new outlet box that has a threaded nipple in its center or installing a mounting strap with a nipple. In either case, the fixture is pushed up over the nipple and a cap is screwed on to hold the fixture in place.

A chandelier that weighs over 10 pounds (4.5 kilograms) requires an extra little hanger called a hickey. It is a strap of metal bent into a U shape. Threaded holes have been made in each leg of the U. One leg is threaded onto the nipple of the box. Another nipple is threaded onto the other leg. The part of the chandelier which will cover the box (called the canopy) will have a hole in it. Shove the

canopy over the nipple. Fasten it in place with the threaded cap provided with the chandelier. See the illustration for a better understanding.

Replacing Receptacles

The receptacle is nothing but a mechanical means of connecting to the circuit. Spring metal contacts accept the prongs of a plug. When this spring metal becomes fatigued and no longer springs back in shape, the plug fits loosely in the receptacle. That is when lights flicker, appliances start and stop, or things simply do not work.

Your new receptacle has a green screw for a grounding wire connection. Modern home circuits contain an extra bare grounding wire (sometimes green or green with yellow stripes if it is insulated) which connects to the green screw and also to the box. If your wiring is old and does not contain the grounding wire, you might still want to connect a short jumper wire from the green screw to a screw or

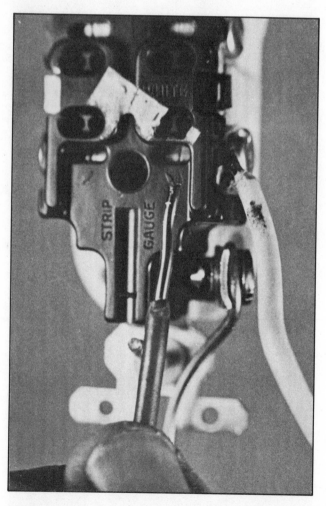

When replacing receptacles, the new types can be connected by either fastening the wires under the screws or by just shoving them into the proper holes in the receptacle. The strip gauge shows you how far back to remove the insulation from the wires.

One hole is marked white. Obviously, the white wire goes into that hole and the black in the other. Tension fittings in the holes grip the wires and make electrical contact.

ground clip on the box. If two conductor, non-metallic shielded cable is used, a separate ground wire would have to be connected to the ground or a metal water pipe. You will at least have the assurance of knowing that when you plug in a drill, for example, a bare hot wire touching the box will not place 120 volts between the box and the frame of the drill you are holding.

If you have difficulty pushing the new receptacle into the box, chances are you straightened out the wires while removing the old one. The wires resist bending and being squeezed into the box. Bend the wires into a loose "S" shape. Now as you push the receptacle in place, the wires offer little resistance as they bend into tighter "S's."

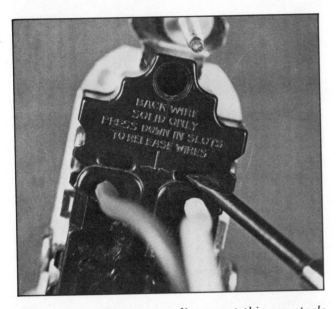

If it becomes necessary to disconnect this receptacle, simply push a small screwdriver into the slots by the holes, press down to relieve the grip, and pull out the wires.

When replacing either receptacles or switches, bend the wires into a loose "S" shape. This will make the wires fold into the box much easier when the switch or receptacle is shoved into position.

Fluorescent Failures

Many persons who have had incandescent lights all their lives have only recently started changing to fluorescents to save on their utility bills. When a fluorescent light stops burning, it is natural to assume that, like an incandescent lamp, the bulb has burned out. Not necessarily. The trouble could also be the starter. It pays to have an extra on hand to try because they do break down often. If neither a new bulb nor a starter gets the lamp working, and you are certain both are making good contact in their sockets, the trouble could be the ballast.

The ballast is a black box inside the fixture. Unless you have a fairly expensive fixture, it will be cheaper to buy a new fixture than have an electrician replace the ballast. Even if you do the work yourself, the replacement ballast will cost almost as much as the fixture.

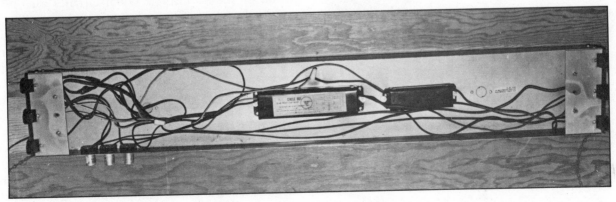

This photo shows the entire inside of a fluorescent fixture. Access was gained by removing the bulbs and then the reflector.

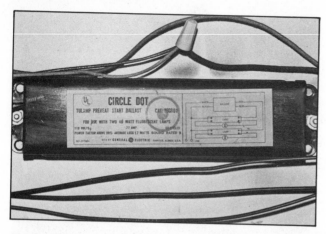

If new bulbs and starters do not get the fluorescent fixture working, chances are a ballast is bad. Lots of ballasts give advance warning by humming loudly for some time before they go bad.

Another frequent surprise to new fluorescent users is a strange buzz in the radio. The radio may be picking up radiation from the fixture either through the air or over the power lines. Try moving the radio into another room. If that does not help, it is getting the radiation from the 120 volt line. Buy a "line filter" from your local TV/radio shop or from a firm such as Radio Shack. The line filter is plugged into a receptacle. The appliance, in turn, plugs into the line filter.

Flickering can be caused by bulbs or starters beginning to break down, or by being loose in their sockets. This also could happen if the fixture has been installed in a cold location (under 50 degrees Fahrenheit, 10 degrees C).

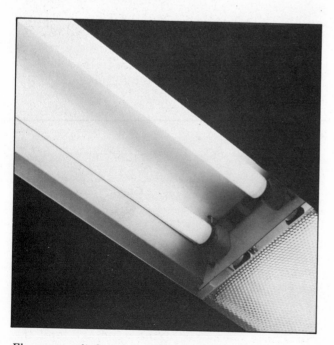

Fluorescent lighting is more economical than incandescent lighting when properly installed.

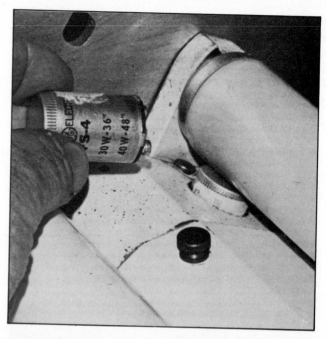

Most fluorescent failures occur in either the bulbs or starters. Bulbs are removed by twisting them 90 degrees counterclockwise, then pulling them free from the sockets at both ends. The top of a starter shows here, just beneath the end of the bulb. Starters also may be on the side of a fixture, away from the bulbs. A new starter is shown so you can see what they look like out of their sockets. They also twist 90 degrees counterclockwise and lift out. To insert a new one, place it in the socket, press down, and twist 90 degrees clockwise.

All fluorescent fixtures hum a bit. It is produced by the ballast. If the hum becomes excessive, the ballast is going bad and may need replacement.

Kitchen Appliance Problems

Plugs and cords on kitchen appliances are generally larger, tougher, and less subject to breakdown than those on lamps, TV's, radios, etc. Nevertheless, old cords do become brittle and insulation falls off the wires. The wires short together or to the frame of the appliance. Open circuits also occur, especially where the cord joins the plug. Troubleshoot these problems with the continuity tester. Use it in the same manner described earlier for checking plugs and cords on lamps. Certainly, the plug and cord combination is the first thing to suspect if the appliance will not work at all or works intermittently. When replacing cords on toasters, irons or similar heating devices, special heating cords must be used. Conventional cords cannot handle the heat.

Plug the appliance into a different receptacle to be sure the trouble is in the appliance before you begin to disassemble it.

If the cord and plug are functioning properly, but

In this side view, you can see the three heating elements which toast the bread.

This is the inside of a small toaster with the chrome cover removed. On the near end is the switching mechanism.

This close-up of the switching mechanism gives you a better look at the contacts. If the points are burned, polish them with an emery cloth. If the spring metal parts are badly bent, it might be possible to reshape them.

the appliance will not start, look for a switch—any kind of switch. On a toaster, the switch engages when the lever is pushed down. A blender has a switch and speed control. The can opener starts when the lever is depressed. A switch mechanism is always suspect in any kind of appliance that has one. Disassemble the appliance. Find the switch, and examine it for burned contacts. Clean them with solvent. Do not oil. If the contacts are not accessible, test the switch with the continuity tester to make sure the contacts are closing. If not, replace the switch.

Appliances such as toasters and hair dryers have heating elements which may burn open or break from fatigue caused by going from cold to hot and back to cold one time too many. Most of these elements are visible when the appliance is disassembled. They look like, and are, lengths of loosely coiled wire. It is a special high resistance wire, however. If you see a break in one, do not attempt to replace it with ordinary wire. Get the correct element from an appliance shop or from the manufacturer.

Electrical troubles are not difficult to locate if you will always keep in mind that an electrical circuit only functions when that circuit is complete. The current must go in one wire of the cord, go through the appliance, and back out the other wire. If the circuit is interrupted (open), the appliance will not work. If a breakdown provides a short-cut to short-circuit the current away from part of the appliance, it will not work either.

Motor Problems

Lubrication may not be an electrical problem, but nevertheless it is the cause of many electric motor troubles. We do not remember to oil our motors. And if we do, we grab a can of thin so-called household oil. A few drops of that on furnace motor bushings, for example, does get it lubricated, but only temporarily. This light oil dries up through use, and it leaves a sticky sort of varnish on the bushings. Before long, the motor cannot fight its normal load and the friction of the varnish, too. It overheats, acrid smoke pours out of the windings, and the motor must be replaced.

A few drops of automobile oil would have pre-

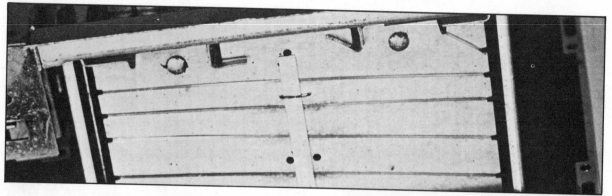

This close-up shows the heating elements are made of long, flat ribbons of resistance wire. If one is burned open, there will be no heat, of course. It will be necessary to replace and rewind the element.

vented this breakdown. Manufacturers generally recommend 20 or 30 weight oil, but any weight of engine oil is better in a motor than household oil. Use the household oil as intended in such things as guns and sewing machine mechanisms.

Not all motors can be oiled. Because people are negligent about oiling motors, some manufacturers use those with sealed bearings. But take a look for yourself. See if there is an oil hole. Check your manual to see if it suggests oiling. And if you were not given a manual, resolve the next time you buy anything that can break down, you will ask for a manual.

The motors in some appliances are not easily accessible, but do not let that stop you. It is much easier and cheaper to disassemble the appliance enough to oil the motor than it is to replace the motor.

Some appliances, such as hand drills and vacuum cleaners, use so-called "universal" motors that operate on either alternating or direct current. These motors are different from others in that they have carbon brushes which rub on the metal segments of a commutator, a device through which current flows to the motor. If the brushes become worn, you will begin to notice sparking. Replace them.

Always clean carbon particles off the commutator before starting the motor. And make sure the commutator bars or segments are not badly discolored from arcing. If they are, it may be necessary to polish them with a non-conducting abrasive like flint paper so they make good contact with the carbon brush.

Anytime you work on a motor, even to oil it, clean accumulated lint and dust away from its ventilating holes. Overheating can break down the insulation in the winding and burn up the motor.

The commutator in a universal motor (this one is in an electric hand drill) is the segmented part around the armature. It is in the center of this picture. The carbon brushes which make contact with the commutator are to the left and right of it. If they are causing sparks, clean the commutator and replace the brushes.

If desired, "in-line" switches can be installed in the middle of electric cords. They often have been used with window fans. This shows the inside connections of one which also has a receptacle at one end. In this case, only the switch was wanted, and the receptacle is not hooked up.

Increasing Efficiency, Reducing Costs

About 80 percent of our electricity is generated by oil, gas, and coal fired turbines. Some 15 percent is generated by hydro-electric. Oil and gas are in short supply.

We have been greedily using more than our share of the world's energy resources in a wasteful manner. Today, we have a great incentive to change all this. It will mean money in our pockets and the possible end to dependence on foreign fuel.

Seven appliances account for 78.2 percent of all the energy consumed in our homes. An equivalent of 2.3 billion barrels of oil per year is used to generate electricity to operate our furnaces, air conditioners, water heaters, dishwashers, laundry appliances, refrigerators, and freezers. Right now the U.S. Department of Energy wants to reduce this use by 20 percent through a long-term plan to make more energy-efficient appliances available for replacement when our old ones wear out.

From this moment on, any time you are in the market for these seven appliances, look for the "Energy Guide" label. It shows the energy-efficiency rating and approximately what it will cost per year to operate. The label lists the lowest and highest energy costs of various models, and how the model compares to others like it. It may cost more to buy energy-efficient appliances, but at today's energy rates, cheap energy-consuming models are more expensive in the long run.

Be careful of the energy costs per year indicated on these labels, however. Currently, they list costs based on rates varying from 2 cents to 12 cents per kilowatt hour. The temptation for the salesman and buyer is to glance at the lowest costs. I do not know where you can buy electricity for 2 cents to 4 cents today. Find out your true rate before shopping. If the rate is not on your electric bills, ask the power company. Or divide any of your bills by the number of kilowatt hours used as indicated on that bill. This will give you the price per kilowatt hour.

If you have to be content with your less energy-efficient appliances there are other ways to reduce energy costs.

Heating

The furnace accounts for roughly 70 percent of your annual fuel bill. Not all furnaces derive their heat from electricity, but general heat conservation methods apply to all types.

A ⅛-inch (3.18 mm) crack around an ordinary size

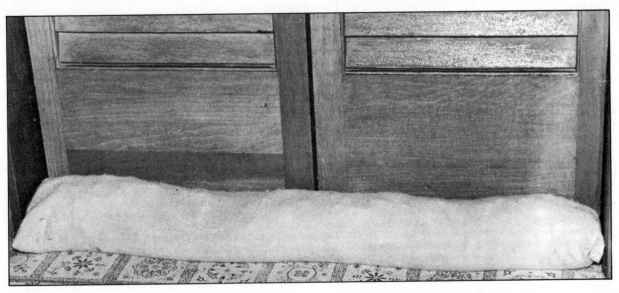

If you feel a draft, heat is blowing out of your house. Use weather stripping plus these portable draft stoppers wherever needed. They are easily made. Sew a piece of cloth in the shape of a tube and fill it with styrofoam nuggets used for packing. Then pour dry sand into the tube. It will fill in around the pieces of styrofoam. Fill the tube to the top. Sew the end shut. Place the stopper on the floor in front of the door where air is blowing in or out.

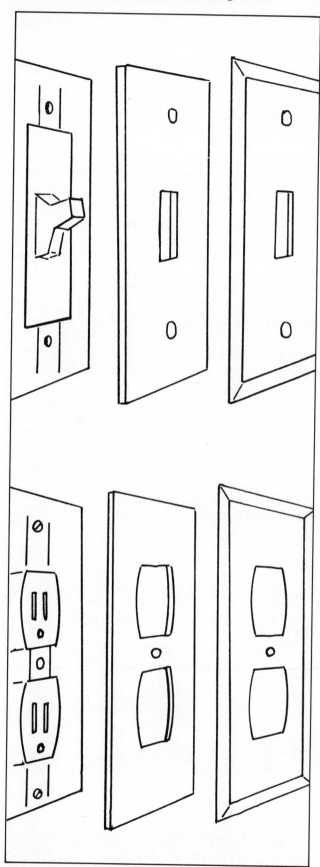

Electrical outlets cause as much as 20 percent of the drafts in our homes. This can be reduced by installing gaskets between switches or receptacles and the cover plates.

door loses heat through 29 square inches (187 square cm). My rural power company estimates this loss at $14 each year.

If you feel a draft, check where it is coming from. Usually, it is coming in one door or window and going out another. The installation of weather stripping around doors and caulking around leaking window frames can eliminate about 30 percent of your home's heat losses.

Storm windows, if you do not already have them, can reduce your heat losses by another 50 percent. They will pay for themselves through energy savings in about 5 years and return 19 percent on the investment every year thereafter.

If you cannot afford storm windows, or are renting, clear plastic can be taped to window frames with transparent weather strip tape to help reduce heat loss. In fact, storm windows have joints which pass cold air. Plastic on the inside of the window makes a third and tighter seal against cold air. Drapes and window shades pulled at night also are a great help toward reducing drafts.

The attic door is a heat waster that is often forgotten. But heat rises. If the door is not tight, heat is lost through the attic. The average home may easily have 30 square inches (approximately 194 square cm) leakage to the attic, raising heat costs about 10 percent.

If there is no insulation in the attic, this is your big loser. Even in mild winters (45 degrees Fahrenheit, 7.2 degrees C), an insulation investment could be returned in two years. You should have at least 6 inches (152.4 mm) of insulation in the attic. An R-30 insulation factor is recommended. R-13 is suggested for walls and in floors over unheated areas. A reasonable amount of insulation can save you 50 percent in heating bills.

Heat savings will not be realized if the furnace is operating inefficiently. Keeping the unit properly serviced can save 10 percent of the heating energy. Clean or replace the filter every four to six weeks. Check the ducts for leaks, and patch with duct tape. Insulate the ducts in unheated areas. Oil the motor bushings (if the motor does not have sealed bearings) twice a year. Make sure the thermostat is on an inside wall and not in a draft.

You also might consider electronic ignition if you have a gas furnace. It eliminates the pilot light. Or, if you have a poor memory, a clogged-filter indicator that tells you by a remote-system light that a dirty filter is reducing heat efficiency.

Turning the heat down at night by only 5 degrees for eight hours can save up to 15 percent of heating energy costs. Ten degrees is much better and is not uncomfortable if you wear pajamas and sleep between sheet blankets. If you have trouble remembering to turn down the heat, you can buy an automatic

Appliance Annual Operation Costs

Appliance	Median Wattage	Median Hours Used Per Year	Annual KWH	Estimated Cost in Dollars Per Year when a KWH is:						
				4¢	5¢	6¢	7¢	8¢	9¢	10¢
Room air conditioner	875	1,000	875	35	43.7	52.5	61	70	79	87.5
Blanket	175	840	147	5.9	7.4	8.8	10	11.8	13.2	14.7
Blender	385	40	15	.6	.8	.9	1.1	1.2	1.4	1.5
Broiler	1,500	70	105	4.2	5.3	6.3	7.4	8.4	9.5	10.5
Coffee maker	700	300	210	8.4	10.5	12.6	14.7	16.8	18.9	21
Deep fryer	1,300	60	78	3.1	3.9	4.7	5.5	6.2	7	7.8
Dehumidifier	500	1,450	725	29	36.2	43.5	50.8	58	65.3	72.5
Dishwasher	1,300	120	156	6.2	7.8	9.4	10.9	12.5	14	15.6
Dryer - clothes	4,800	200	960	38.4	48	57.6	67.2	76.8	86.4	96
Fan - window	200	850	170	6.8	8.5	10.2	11.9	13.6	15.3	17
Freezer	450	4,000	1,800	72	90	108	126	144	162	180
Fry pan	1,200	160	192	7.7	9.6	11.5	13.4	15.4	17.3	19.2
Furnace fan	800	1,200	960	38.4	48	57.6	67.2	76.8	86.4	96
Hair dryer	750	50	38	1.5	1.9	2.3	2.7	3	3.4	3.8
Heater - space	1,300	135	176	7	8.8	10.6	12.3	14.1	15.8	17.6
Iron	1,000	145	145	5.8	7.3	8.7	10.1	11.6	13	14.5
Microwave oven	650	135	88	3.5	4.4	5.3	6.2	7	7.9	8.8
Mixer	125	100	13	.5	.6	.8	.9	1.0	1.17	1.3
Radio	75	1,200	90	3.6	4.5	5.4	6.3	7.2	8.1	9
Range	12,000	100	1,200	48	60	72	84	96	108	120
Refrigerator	300	3,500	1,050	42	52.5	63	73.5	84	94.5	105
Rotisserie	1,300	150	195	7.8	9.8	11.7	13.6	15.6	17.5	19.5
Stereo	400	1,000	400	16	20	24	28	32	36	40
TV	250	2,200	550	22	27.5	33	38.5	44	49.5	55
Toaster	1,200	35	42	1.7	2.1	2.5	2.9	3.4	3.8	4.2
Vacuum cleaner	550	75	40	1.6	2	2.4	2.8	3.2	3.6	4
Waffle iron	1,100	20	22	.9	1.1	1.3	1.5	1.8	1.9	2.2
Washer - clothes	800	225	180	7.2	9	10.8	12.6	14.4	16.2	18
Water heater	3,500	1,500	5,250	210	262	315	367	420	472	525

This chart estimates the annual operation costs of most common household appliances. Note that the estimates are based on median wattages and numbers of hours used. Your appliance may not use the same number of watts, and you may use it more or less than the average person. Currently, heating and cooling accounts for 69 percent of the average family's electric bill, refrigeration and cooking 14 percent, hot water 10 percent, small appliances 5 percent, and lighting only 2 percent.

thermostat timer that does the job for you.

Train your family to dress warmly so the thermostat can be kept lower. Sweaters are cheaper than energy. Make sure everybody gets in and out of doors quickly and shuts them tightly on the first try. Lower the thermostat when a large group of people gathers. Seal off unheated rooms. Humidify for better comfort at lower temperatures. Do not block registers with furniture. Do not heat the kitchen with the oven. If you have a fireplace, make sure the damper is closed when not in use. Install glass doors over the fireplace. Heat does not just rise out of fireplace chimneys, it is sucked out by the draft.

Electrical outlets were found to be responsible for as much as 20 percent of the drafts in the house. This can be greatly reduced by adding gaskets under switch and receptacle plates, and by placing plastic inserts in the receptacles. The inserts have the added advantage of possibly preventing a very young child from getting shocked sticking something into a receptacle.

Hot Water Heaters

Heating water, especially with electric hot water heaters, is a real energy waster. It is our number two energy consumer, and we pay dearly for the convenience. Heating water around the clock is a little like letting your car run all the time so it is more convenient to jump in and go.

If you have an electric hot water heater, by all means get a timer. It can shut off the heater during those hours you do not need hot water, hours when perhaps you are not even at home, and turn it on in time for your return. The timer also can shut off the

Air also passes through the slots in receptacles. This can be stopped by filling the unused receptacles with plastic inserts.

heater at night and have hot water ready again by morning.

In some areas, power companies will allow cheaper rates on electricity for water heaters if those heaters are not in operation during peak demand hours. In these cases, the power companies usually install the timer.

Most heaters in use today throw away as much as 14 percent of their heat through poor insulation around the tank. You can buy an insulated blanket to cover the heater that will pay for itself in about a year.

The hotter you keep the water in the tank, the more it will cost you. Reducing the setting from 140 degrees Fahrenheit (60 C) to 120 (48.89 C) will decrease energy consumption by 18 percent.

Baths cost more than showers—about 2,000 gallons (7,570 L) more of hot water per person each year. If you further conserve by adding a flow constricter in the shower head, you can save another $10 per year for every member of the family.

Do not waste hot water. A faucet that leaks one drop per second wastes 650 gallons (2,460.25 L) a year. That is enough for 59 loads of wash. And do not leave water running while you rinse dishes or shave. Insulate long runs of hot water pipes.

Electric hot water heaters should be on timers. This way, you won't be wasting energy by keeping water hot all night or during times when you are not home.

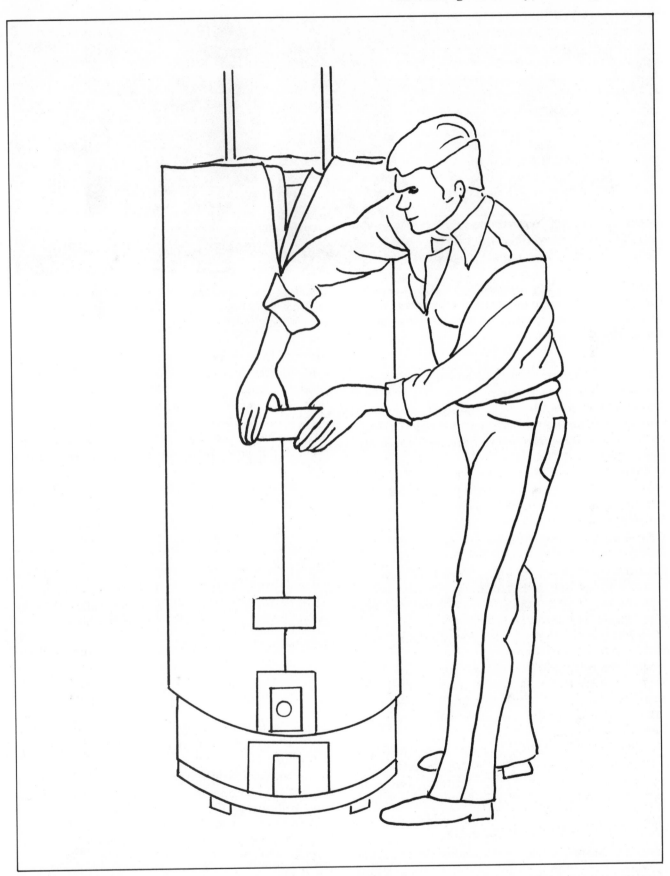

Water heaters lose as much as 14 percent of their heat through poor insulation around the tank. An insulated blanket to cover the heater will pay for itself in about a year.

The glass cooking surface (at left) on this countertop range unit is not hot to the touch. It only heats the pan which in turn cooks the food. Heat is not released into the room. Photo courtesy of Tappan

Cooking

Electrical cooking costs can be cut 50-75 percent by using a microwave oven.

Keep reflectors under the burners of conventional ranges clean so they reflect heat instead of absorbing it. And clean burners radiate heat more efficiently than dirty ones.

Never put a little pan on a big burner. Too much heat would be lost to the air. And the contents come to a boil much faster with less energy in a covered pot than in an open pan.

Burners on electric ranges are slow to heat and slow to give up their heat. Do not keep the burner on until the cooking is finished. With a little practice, you can judge how many minutes ahead you can turn off the burner and still finish cooking. This also applies to baking in the oven.

Do not use the big oven for a small dish. Try to cook as many things as possible at once in the oven. Sometimes refrigerated or frozen dishes that were prepared earlier can be used to make an all-oven meal. And learn how long different dishes require to bake so a timer can be used instead of opening the door to check.

If there is a choice, use the range top instead of the oven. And better yet, use an electric fry pan.

Pressure cookers use much less energy than ordinary covered pots by greatly reducing cooking time.

Refrigeration

Keeping things refrigerated requires 750 to 1,400 kilowatt hours each year, depending on the size and condition of the refrigerator and whether the unit has automatic defrosting. Models with self-defroster freezer compartments are costlier to buy and operate, and in some instances have caused freezer burn and shorter freezer life for stored food. Keep the frost less than ¼-inch (6.35 mm) thick on manual defrosting models. An upright freezer is less efficient than the chest type. Cold air falls, so it tumbles out whenever the door is opened on an upright model. The same principle applies to the door of a refrigerator.

If the door gasket is making a poor seal, it is like leaving the door partially open all the time. Check this by closing the door on a piece of paper. If you can easily pull out the paper, you probably need a new gasket. The same thing applies to an upright freezer door, and to a lesser extent, a chest type.

Take the refrigerator's temperature now and then. An ordinary thermometer will do. It should read between 38 and 42 degrees Fahrenheit (3.33 to 5.55 C). You may have to adjust the control to achieve that range. Insulation is not always what it might be in a refrigerator, so prevailing room temperatures make a big difference in the necessary control setting. It may require resetting four times a year or more.

Freezer compartments should be around 5 degrees Fahrenheit (-15 C). Colder settings waste electricity.

Pull the refrigerator away from the wall and clean the motor housing and condenser coils twice a year. You probably will have to remove a cover at the bottom of the backside of the refrigerator for cleaning. Use a paint brush to brush off the dirt, and have a vacuum cleaner running to suck it in as fast as you stir it up.

Most models (except built-ins) should be kept an inch or two (25.4 to 50.8 mm) from the wall for good air circulation around the compressor.

This well-designed, attractive kitchen is energy efficient as well. The light color and decor of this room do not require much light. A large, overhead light illuminates the cooking and dining island while several small lights in the soffit and beneath the wall cabinets provide special lighting. A microwave oven is used rather than a conventional oven. Photo courtesy of American Olean Tile

Try to keep the refrigerator filled to capacity without blocking air circulation. Full freezers require less energy than those half-empty.

Keep refrigeration units away from ranges or heat registers.

Other Kitchen Tips

Use the dishwasher only when you have a full load. And turn it off after the final rinse. Let the dishes air-dry. If the unit does not have an air-dry switch, shut it off and prop the door open for faster drying. Avoid the "rinse hold" on your dishwasher. It costs you three to seven gallons (11.3 to 26.4 L) of water each time it is used. And use cold water if you rinse the dishes before loading them into the dishwasher. Models with air-power and overnight dry settings save about 10 percent in dishwashing energy costs.

Buy an aerator for the kitchen faucet. It reduces the water flow and therefore the energy to heat that water.

Lots of people think hot water is needed to wash greasy garbage down the disposal. Quite the opposite is true. Cold water solidifies grease, making it easier to be ground up with the other garbage and washed away instead of clinging to pipes.

Lighting around the bathroom lavatory is important. An overhead light and two side lamps provide uniform light for shaving and makeup application. Such lighting illuminates the face without causing shadows. Photo courtesy of Armstrong Cork Co.

Air Conditioning

Choose the right size equipment. Too large or too small a unit is inefficient. Locate the unit on a shady side of the house, away from shrubs and the dryer vent. Clean filters regularly. Do not obstruct window units with drapes or let furniture get in the way of air circulation. Do not run a window fan when the air conditioner is working. Keep the thermostat out of sunlight and away from TVs or any other heat-producing appliance. Close shades on sunny windows. An attic ventilating fan expels hot air, reducing the air conditioner's work load. Plant trees or use awnings to block the sun. Close all floor and wall registers to keep cold air from falling through. Keep the basement door closed.

Lighting

It is cheaper to burn one 100 watt bulb than two 60's, and you get more light. But most of the energy used by incandescent bulbs goes up in heat. Choose fluorescent if possible.

If you are stuck with incandescents, avoid tinted bulbs. And stay away from "long life" bulbs. They are designed for higher voltage than 120, so when operated at 120, they last longer, but use more current for less light the entire time.

Paint ceilings and walls a light color to bounce light around the room instead of absorbing it. And keep bulbs and lamp shades clean for maximum light radiation.

Install a dimmer when bright light is unnecessary or undesirable. And when you leave the room, turn out the light.

If you are planning future lighting changes, consider the relative efficiency of the available choices. Incandescent bulbs deliver 8 to 22 lumens of light for every watt they consume, almost 90 percent of the energy going up in heat. Fluorescents give 30 to 83 lumens per watt. And for yard lights and outbuildings, mercury provides 26 to 58 lumens, metal halide, 67 to 115, and high pressure sodium, 74 to 132.

Laundry

Wash with cold water whenever possible. Always use cold water in the rinse cycle. Match the water level with the load size. Presoaking allows shorter wash cycles. Do not overload the machine, but do not underload, either. Use the full spin cycle to get the clothes as dry as possible before putting them in the dryer.

Fill, but do not overload the dryer. Clean the lint screen after each load. Overdrying wastes electricity. The dryer cools and must be brought back up to temperature if allowed to stand idle between loads. Try to avoid this by drying continuously, one load after another, until the laundry is finished. Separate the heavy and light clothing, drying the heavy items first. Lightweight garments may partially or completely dry without power from the heat still remaining in the dryer. Remove clothes promptly to avoid unnecessary ironing.

Heat from the dryer should not be piped outside and wasted during the winter months. Of course, you cannot blow lint all over the house either. But the strap can easily be removed to disconnect the flexible hot air pipe from the outside vent. And it can be just as easily reconnected to an inside unit designed to trap lint, yet let the hot air blow into the house.

A second benefit to an inside vent is the added moisture it blows into the house at a time when it is difficult to keep the humidity high enough.

A dryer diverter is available which switches between outside or inside venting without having to remove or reconnect straps. If the dryer is in a heated area, the heat and moisture is said to save as much as $12 a month, about the cost of such a unit.

Glossary

Alternating current (AC) Voltage changes polarity, and current changes direction, 60 times per second in the United States—50 times in some other countries.

Ampacity The maximum current-carrying capacity of a wire.

Ampere The unit of current flow (movement of electrons along a wire). The number of amperes flowing can be determined by dividing the voltage applied (in volts) by the resistance (in ohms) offered by the device which is connected to this voltage.

Ballast Black box device in fluorescent fixtures used to start and maintain light at a steady level.

Branch circuits Circuits branching out from the circuit breaker panel or fuse box.

Bus bar Uninsulated metal bar with numerous screws for connecting neutral and grounding wires in a circuit breaker panel or fuse box.

BX cable A trade name for armored cable.

Cable May be any kind of wire, but in residential wire often used to mean two or more wires in a single sheath.

Circuit breakers Circuit protection device that trips to "off" position when excessive current flows and may be reset again to the "on" position. Unlike a fuse which blows when overloaded and is ruined, the circuit breaker may be reset many times.

Conductors Usually wires, but any metal which conducts electrical current.

Conduit A pipe for enclosing electric wires. It protects the wires in hazardous locations.

Connection Bring together two or more wires or contacts.

Contacts Two or more conducting surfaces which when they touch together allow current to flow, as in switches.

Current The movement of electrons in a wire.

Direct current (DC) Current which flows in only one direction.

Electron Small particle, one or more of which orbits around a nucleus in an atom. A hydrogen atom has one proton in its nucleus around which one electron orbits, much as the moon orbits the earth. Conducting materials have many more electrons in their atoms, and some are free to move from atom to atom, therefore "electron flow."

Ell An L or elbow-shaped conduit fitting with a removable cover to aid in getting wires through a 90 degree bend.

End-of-run outlet Last outlet on a circuit.

Entrance panel The fuse box or circuit breaker to which power enters from the meter.

Fish tape Flexible narrow metal tape designed to pull wires through walls, conduit, etc.

Four-way switch A switch that allows control of a light from three or more locations when used in conjunction with two 3-way switches.

Ground The earth voltage potenial of zero to which one side of a circuit is connected for reasons of safety.

Ground-fault interrupter (GFI) A highly sensitive circuit breaker which trips when the current on the hot side of a circuit is slightly higher than on the neutral. When this happens, it means that a small amount of current is leaking through a high resistance short to the frame of some device, such as a hand drill. The GFI interrupts the circuit before the person holding the drill, or other device, can be severely shocked.

Grounding Connecting to ground by means of rods driven into the earth or water pipes already underground. We are grounding a circuit when we connect the white wire of that circuit to such rods or pipes.

Heating tape A tape containing a heating element. May be used for keeping ice off the edge of roofs, keeping water pipes from freezing, keeping animals warm in winter, etc.

Hickey Metal device used to hang a chandelier on an outlet box. Also a device for bending conduit.

Hot wire The wire which carries the voltage. It should always be the black wire, or if another color must be used, it should be identified as the hot wire by wrapping black tape around it at both ends. Blade and red hot wires are permitted in 3-wire, 240 volt circuits.

Incandescent lighting Light created by heating a wire in a vacuum inside a glass bulb.

Indirect lighting Light that is aimed at walls and/or ceiling and bounces around the room. Diffused, pleasant light, but a very inefficient method.

Insulation Any material which will not conduct electricity. Examples are the plastic and rubber coatings on wire, and electrician's tape.

Insulator Any device which permits current flow.

Junction box Outlet box intended as a place to splice or join wires.

Kilowatt One thousand watts of electrical power.

Kilowatt hour One thousand watts used for one hour. The unit by which the power company charges us for electricity.

Knockout Small, coin-size, partially stamped out circles in outlet boxes, circuit breaker cabinets, etc., that can be easily knocked out to make a hole to admit entry of a cable.

Lumens The unit of measuring the amount of light being emitted from a source such as a light bulb.

Meter There are many types of meters, but in this text we are referring only to the meter by which the power company measures the number of kilowatt hours we use.

Middle-of-run outlet As opposed to end-of-run where circuit terminates, middle-of-run accepts the circuit and passes it along to the next outlet.

Neutral bus bar Metal bar in fuse box or circuit breaker to which all neutral wires are connected.

Neutral wire The wire which is grounded and therefore carries no voltage. It should always be the white wire, and if it is impossible to use a white wire as the neutral, it should be identified as white by wrapping it with white adhesive tape at each end.

Nipple Short length of threaded pipe used in hanging fixtures.

Non-metallic sheathed cable Common name Romex. A cable containing two or more insulated wires in a single plastic sheath.

Ohm The unit of measuring resistance to the flow of electrons.

Outlet Any socket, receptacle, or switch where electricity has been brought out of the circuit in some manner to be put to some use.

Outlet box Box which holds any socket, receptacle, or switch.

Pigtail socket Rubber-coated with two wires molded in.

Pigtail splice Simple manner of twisting two wires together to join or splice them.

Plug Any one of a variety of pronged devices used to connect a cord into a receptacle.

Raceway Channels and outlets for a type of surface wiring.

Receptacle Outlet providing voltage for any electrical device designed to be plugged in.

Resistance Resistance to the flow of electrons caused by the nature of the material. The atoms of some materials give up and pass along free electrons easily (example: copper) while others are reluctant to give up electrons (example: resistance wire in toasters) and generate heat in the process. All conductors offer some resistance to current flow.

Romex The trade name for non-metallic sheathed cable.

Service entrance Where power company wires enter the house.

Service head Another name for weatherhead, a part of the service entrance which accepts the power company lines.

Service panel Part of a circuit breaker cabinet or fuse box which accepts the power company voltage and distributes it in branch circuits throughout the house.

Short circuit A breakdown of insulation which allows current to short-cut its normal path of flow.

Snaking An alternate term for fishing (pulling) wire through a wall.

Splicing Joining two or more wires.

Terminals Connections such as the screws on receptacles, switches, etc., which join wires to a device and hold them in place.

Thinwall Readily bent metal conduit for housing and protecting cables in hazardous locations.

Three-way switches Switches designed to control a light from two locations.

Transformer A device with primary and secondary coils for changing voltage or current up or down. A small one is used to change 120 volts to 10 or 16 volts for door bells. A large one on a power company pole delivers 240 to 120 volts to our homes.

Underground feeder (UF) A type of cable encased in solid, but flexible, plastic which permits it to run underground without danger of deterioration from the soil's acids.

Underwriter's knot Knot that protects the connection in a plug from pulling loose.

Underwriters' Laboratories, Inc. Organization that tests electrical devices and gives them a "UL" label if safe.

Volt Unit of measuring electrical pressure.

Voltage Amount of volts at a certain point or place.

Watt Unit of electrical power.

Weatherhead That part of an above ground service entrance which accepts the power company wires.

Index

Metric Conversion Tables

Length Conversions

fractional inch	millimeters	fractional inch	millimeters
1/32	.7938	17/32	13.49
1/16	1.588	9/16	14.29
3/32	2.381	19/32	15.08
1/8	3.175	5/8	15.88
5/32	3.969	21/32	16.67
3/16	4.763	11/16	17.46
7/32	5.556	23/32	18.26
1/4	6.350	3/4	19.05
9/32	7.144	25/32	19.84
5/16	7.938	13/16	20.64
11/32	8.731	27/32	21.43
3/8	9.525	7/8	22.23
13/32	10.32	29/32	23.02
7/16	11.11	15/16	23.81
15/32	11.91	31/32	24.61
1/2	12.70	1	25.40

feet	meters	feet	meters
1	.3048	8	2.438
1½	.4572	8½	2.591
2	.6096	9	2.743
2½	.7620	9½	2.896
3	.9144	10	3.048
3½	1.067	10½	3.200
4	1.219	11	3.353
4½	1.372	11½	3.505
5	1.524	12	3.658
5½	1.676	15	4.572
6	1.829	20	6.096
6½	1.981	25	7.620
7	2.133	50	15.24
7½	2.286	100	30.48

inches	centimeters	inches	centimeters
1	2.54	5	12.70
1¼	3.175	5¼	13.34
1½	3.81	5½	13.97
1¾	4.445	5¾	14.61
2	5.08	6	15.24
2¼	5.715	6½	16.51
2½	6.35	7	17.78
2¾	6.985	7½	19.05
3	7.62	8	20.32
3¼	8.255	8½	21.59
3½	8.89	9	22.86
3¾	9.525	9½	24.13
4	10.16	10	25.40
4¼	10.80	10½	26.67
4½	11.43	11	27.94
4¾	12.07	11½	29.21

Common Conversion Factors

	Given the number of	To obtain the number of	Multiply by
Length	inches	centimeters (cm)	2.54
	feet	decimeters (dm)	3.05
	yards	meters (m)	0.91
	miles	kilometers (km)	1.61
	millimeters (mm)	inches	0.039
	centimeters	inches	0.39
	meters	yards	1.09
	kilometers	miles	0.62
Area	square inches	square centimeters (cm²)	6.45
	square feet	square meters (m²)	0.093
	square yards	square meters	0.84
	square miles	square kilometers (km²)	2.59
	acres	hectares (ha)	0.40
	square centimeters	square inches	0.16
	square meters	square yards	1.20
	square kilometers	square miles	0.39
	hectares	acres	2.47
Mass or weight	grains	milligrams (mg)	64.8
	ounces	grams (g)	28.3
	pounds	kilograms (kg)	0.45
	short tons	megagrams (metric tons)	0.91
	milligrams	grains	0.015
	grams	ounces	0.035
	kilograms	pounds	2.21
	megagrams	short tons	1.10
Capacity or volume	fluid ounces	milliliter (ml)	29.8
	pints (fluid)	liters (l)	0.47
	quarts (fluid)	liters	0.95
	gallons (fluid)	liters	3.80
	cubic inches	cubic centimeters (cm³)	16.4
	cubic feet	cubic meters (m³)	0.028
	cubic feet	liters	28.3
	bushels (dry)	liters	35.2
	milliliters	ounces	0.034
	liters	pints	2.11
	liters	quarts	1.06
	liters	gallons	0.26
	liters	cubic feet	0.035
	cubic centimeters	cubic inches	0.061
	cubic meters	cubic feet	35.3
	cubic meters	bushels	28.4
Temperature	degrees Fahrenheit	degrees Celsius	0.556 (after subtracting 32)
	degrees Celsius	degrees Fahrenheit	1.80 (then add 32)